D0206482

DIGITAL
TEXTILE
PRINTING

Textiles that Changed the World

Series Editor: Linda Welters, University of Rhode Island, USA

ISSN: 1477-6294

Textiles have had a profound impact on the world in a multitude of ways – from the global economy to the practical and aesthetic properties that subtly shape our everyday lives. Pioneering in approach, this series chronicles the cultural life of individual textiles through sustained, book-length examinations, focusing on historical, social and cultural issues, and the myriad ways in which textiles ramify meaning. Each book is devoted to an individual textile, to a dye, such as indigo or madder, or to a technique that characterizes a particular type of cloth. Books in the series are handsomely illustrated with color as well as black-and-white photographs.

Previously published in the Series

Tartan
Jonathan Faiers

Felt
Willow Mullins

Cotton
Beverly Lemire

DIGITAL TEXTILE PRINTING

SUSAN CARDEN

Bloomsbury Academic
An imprint of Bloomsbury Publishing Plc

B L O O M S B U R Y
LONDON • NEW DELHI • NEW YORK • SYDNEY

LONDON PUBLIC LIBRARY

Bloomsbury Academic

An imprint of Bloomsbury Publishing Plc

50 Bedford Square   1385 Broadway
London   New York
WC1B 3DP   NY 10018
UK   USA

www.bloomsbury.com

**BLOOMSBURY and the Diana logo are trademarks of
Bloomsbury Publishing Plc**

First published 2016

© Susan Carden, 2016

Susan Carden has asserted her right under the Copyright, Designs and
Patents Act, 1988, to be identified as Author of this work.

All rights reserved. No part of this publication may be reproduced or
transmitted in any form or by any means, electronic or mechanical,
including photocopying, recording, or any information storage or retrieval
system, without prior permission in writing from the publishers.

No responsibility for loss caused to any individual or organization acting
on or refraining from action as a result of the material in this publication
can be accepted by Bloomsbury or the author.

British Library Cataloguing-in-Publication Data
A catalogue record for this book is available from the British Library.

ISBN: HB: 978-1-4725-3568-9
       PB: 978-1-4725-3567-2
       ePDF: 978-1-4742-6029-9
       ePub: 978-1-4742-6028-2

Library of Congress Cataloging-in-Publication Data
Carden, Susan.
Digital textile printing / by Susan Carden.
pages cm. -- (Textiles that changed the world, ISSN 1477-6294)
Includes bibliographical references.
ISBN 978-1-4725-3568-9 (HB) -- ISBN 978-1-4725-3567-2 (PB) --
ISBN 978-1-4742-6029-9 (ePDF) -- ISBN 978-1-4742-6028-2 (ePub)
1. Textile printing. 2. Digital printing. I. Title.
TP390.C37 2015
667'.38--dc23
2015016055

Series: Textiles that Changed the World, 1477-6294

Typeset by Fakenham Prepress Solutions, Fakenham, Norfolk NR21 8NN
Printed and bound in India

CONTENTS

List of Plates   VII
List of Figures   IX
Foreword   XI
Acknowledgments   XIII

1   Introduction   1
2   History of Printed Textiles   7
3   Technologies, Substrates and Dyes   23
4   The Process of Digital Textile Printing   35
5   Art and Design Practice   51
6   Essence of Digital Printing   71
7   Investigating Digital Textile Printing   81
8   Crossing Disciplines   93
9   Conclusions   111

Glossary   117
Bibliography   127
Index   141

PLATES

Front cover: Digitally printed silk satin. Susan Carden. *Digital textile print*. 2013. Photograph by Louis Carden.

Plate 1   Multilayered digital print. Susan Carden. Multilayered digital print. 2013. Photograph by Susan Carden.

Plate 2   Detail hand stencil. Marcos García-Diez. Hand stencil El Castillo. 2015. University of the Basque Country. Photograph by Marcos García-Diez.

Plate 3   Bed curtain and valance of plate-printed cotton, printed and made by Nixon & Co, Great Britain, 1770–1780. Bed curtain and valance. 2015. © Victoria and Albert Museum, London.

Plate 4   Furnishing fabric with a design of fantastic flowers with pencilled blue. Furnishing fabric. 2015. © Victoria and Albert Museum, London.

Plate 5   Letterpress. Christopher Wakeling. *Letterform 1*. 2013. Photograph by Susan Carden.

Plate 6   Furnishing fabric of roller-printed cotton, Lancashire, 1831. Furnishing fabric. 2015. © Victoria and Albert Museum, London.

Plate 7   Digital image on silk. Susan Carden. Silkscreen print. 2010. Photograph by Susan Carden.

Plate 8   Generated digital image. Alex Russell. *Cloth of Gold 1*. © Alex Russell 2013.

Plate 9   Laser digital print. Kate Goldsworthy. *Mono finishing-01*. 2009. Photograph by Kate Goldsworthy.

Plate 10   Make-ready sample in full color subsequently divided into the cyan layer and the magenta layer. Steve Rigley. 2007. Photograph by Steve Rigley.

Plate 11   Digital textile print. Steve Rigley. Sivakasi Textile Sample 1. 2007. *Journal of Craft Research* 4 (2). Photograph by Steve Rigley.

Plate 12   Digital textile print. Steve Rigley. Sivakasi Textile Sample 2. 2007. *Journal of Craft Research* 4 (2). Photograph by Steve Rigley.

Plate 13   Textile print. Robert Stewart. *Macrahanish*. (1954) Digitally reproduced 2003.

Plate 14   Textile print. Robert Stewart. *Ardentinny*. (1954–55). Digitally reproduced 2003.

Plate 15   Textile design pencil and watercolor on paper 29.2 × 20.9. Charles Rennie Mackintosh. Stylized Chrysanthemums. © The Hunterian, University of Glasgow. 2015.

Plate 16   Digital textile print. Helena Britt. Assembly. 2009. Photograph by Helena Britt and Elaine Bremner.

Plate 17   Digital textile print. Elaine Bremner. Awaken Collection. 2009. Photograph by Helena Britt and Elaine Bremner.

Plate 18   Silkscreen-print. Susan Carden. Silk digital textile print. 2010. Photograph Susan Carden.

Plate 19   Illustration 22: Digital textile print on pleated silk. Susan Carden. Craft stitch technique. 2010. Photograph Susan Carden.

Plate 20   Digital textile print on silk. Susan Carden. Craft technique 2. 2010. Photograph Susan Carden.

Plate 21   Silk digital print. Susan Carden. Digital ikat on silk dupion. 2013. Photograph Susan Carden.

Plate 22   3D printed teddy bear. Scott E. Hudson. 2014. Carnegie Mellon University.

Plate 23   Print head detail, Scott E. Hudson. 2014. Carnegie Mellon University.

Plate 24   Silk textile print. Susan Carden. Silk dupion digital print. 2010. Photograph Susan Carden.

Plate 25   Silk digital print. Susan Carden. Pleated resist image. 2011. Photograph Susan Carden.

Plate 26   Silk satin digital print. Susan Carden. Transfer and batik image. 2011. Photograph Susan Carden.

Plate 27   Silk satin digital print. Susan Carden. Batik transfer image. 2011. Photograph Susan Carden.

Plate 28   Stencil image on silk. Susan Carden. Stencil image on silk. 2011. Photograph Susan Carden.

Plate 29   Batik and transfer image on silk. Susan Carden. Batik and transfer image on silk. 2011. Photograph Susan Carden.

FIGURES

1 Stages of digital textile printing. Susan Carden 2010. 35

2 3D digital print. Cathy Treadaway. *Sennen*. 2011. Photograph by Dirk Dahmer. 55

3 Laser digital print. Kate Goldsworthy. *Laser line*. 2013. Photograph by Kate Goldsworthy. 57

4 Laser digital print. Kate Goldsworthy. *Mono finishing-02*. 2009. Photograph by Kate Goldsworthy. 58

5 Feed head cut away. Scott E. Hudson 2014. Carnegie Mellon University. 67

6 Knowledge in digital textile printing. Susan Carden 2013. 74

7 Digital textile printing as research. Susan Carden 2012. 83

8 Post-consumer waste: paper to print on fabric. Susan Carden 2011. 91

9 Textile design and neighboring disciplines. Susan Carden 2011. 93

10 Visibility of digitally printed images on silk. Susan Carden 2012. 108

11 Development of a digital textile print. Susan Carden 2011. 110

FOREWORD

This book is an inquiry into the nature, meaning and possibilities of digital textile printing from the perspective of a practising designer, academic and researcher. As this is a practice-based position, it facilitates a view across the application of advanced technology combined with creating and making by hand. It is primarily an exploration of how digital textile printing evolved, establishes its defining factors, and reflects on the wide range of new opportunities for contemporary designers and artists. Also highlighted are a number of ways in which practitioners are maximizing the advantages of advanced technology.

Digital textile printing is currently located at an interesting junction: it is inextricably linked to advances in technology, while at the same time it requires materials that originate in nature, visual creativity which is abstract, and, increasingly, it is produced within a demonstrably sustainable context. Not only are designers and their clients often located on different continents, but the various processes and components of digital textile printing may be situated in another place altogether.

This book also looks at how designers are approaching and challenging the conditions of practice, including associated opportunities. It shows that what may seem like a closed black-box system, in which a practitioner cannot influence how the process of digital printing is carried out, is not always the case. Even a limited number of opportunities, such as intervening after the coded image has arrived at the printer but before the dye has been applied, can offer scope for personal expression that results in novel outcomes.

The author recognizes and considers the pioneering work of a number of experienced practitioners and developers in this field, and reflects on the outcomes of a series of interviews conducted with designers who regularly use advanced technology in their work. These responses have helped to define multiple aspects of the digital textile printing provision, and while some characteristics impact positively on the images or motifs, other factors can nudge the design process down a path that is not always desirable. All of these positions are highlighted and presented from a variety of perspectives so that the reader can see the arguments from different sides. However, all the practitioners interviewed recognize the importance of advanced technology and are eager to utilize selected features of the process within their own practice.

The aim of this book is to demonstrate what can be achieved with digital textile printing. It also provides examples, and explanations, of how a variety of practitioners have successfully managed to negotiate their way through the range of options available to them. Creativity and the urge to make by hand are well embedded in the human condition, with handcrafted textile production of clothing and shelter being essential to humankind's survival as a species. When the opportunity is available to designers to combine these well-established skills with technological systems and processes, situations most practitioners would not previously have had to deal with, there is an opportunity to create something unexpected and extraordinary. Throughout this book it is the intention to better understand where the desire to impart marks on a movable surface began, and see where they might be located in a contemporary situation. These impulses started with painting on the walls of caves, but have now evolved to become compatible with creating images and patterns that are formed as code. This volume also seeks to identify how this creativity has been embraced and expanded upon in the digital landscape. This new environment encourages everyone to create more quickly, produce more, think about the pressure on natural resources, consider the methods and means of disposing of an item before we start to make it, and to do all of this to the highest possible aesthetic standards. These are ambitious goals for textile practitioners, but by examining the resources available to them, this book attempts to enable designers to create more efficiently and, therefore, more effectively.

ACKNOWLEDGMENTS

This book would not have been possible without the kind support of a number of people. First, I would like to thank Professor Linda Welters, Geraldine Billingham, Agnes Upshall and Hannah Crump for their kind advice, recommendations and help with realizing this publication. Second, from the Glasgow School of Art, I would like to thank Professor Irene McAra McWilliam, Jimmy Stephen-Cran, Dr Nicky Bird, Dr Laura Gonzalez, and Professor Thomas Joshua Cooper, as well as Alan Shaw and Vicky Begg from the Centre for Advanced Textiles at GSA for all their support and enthusiasm. Third, I am most grateful to Christopher Wakeling, Dr Marcos García-Diez, Alex Russell, Dr Cathy Treadaway, Dr Kate Goldsworthy, Professor Scott Hudson, Steve Rigley, Dr Helena Britt and Elaine Bremner for kindly granting me permission to reproduce their images and photographs in this book. I would particularly like to thank Dr Sheila Stewart for generously allowing me to use a number of textile images designed by her late husband, Robert Stewart. Finally, but by no means least, I would like to thank my family for being the biggest inspiration and support of all.

1

INTRODUCTION

The arrival of digital textile printing at the end of the twentieth century has had a profound effect on the nature of textile design. The significance of this new technology, coupled with advanced developments in fabric and dye chemistry, is only now starting to become clear. This volume explains how the technology used in digital textile printing evolved and why advances in one area of art and design practice have influenced many neighboring fields. This book also considers the risks to creative individuality brought about by such a rapid digital expansion. Through an exploration of reflective studio inquiry, an argument is presented for clarifying the boundaries of authenticity within textile design. In order to support future developments, rigorous terms of reference are provided to inform decision-making within the studio environment, and for evaluating the status of digitally printed textile design across the wider academic community.

Chapter 1 looks at the history of textile printing. Printing is one of the earliest forms of artistic mechanical reproduction. Empirical evidence shows that the Ancient Greeks were able to replicate an image in two distinct ways: stamping, in which an image or motif is applied to a surface through pressure contact from a raised pre-formed pattern shape, such as carved wood, and founding, a process that pours a molten material, for example metal or glass, into a prepared mold of clay or sand (Benjamin 1936; Updike 1962; Harthan 1983; Laucy 2010). Printing evolved from these two basic creative processes that allow reproductions to be made. However, by the seventeenth century, India led the world in textile printing due to its superior knowledge of mordants and dyeing techniques. The Europeans, lacking this information, were unable to compete with, for example, the bright purple-blue of indigo and the vibrant reds obtained from madder grown in the calcium-rich soil of Southern India. In comparison, European fabrics tended to fade easily, the dye ran when washed, and when trade with India blossomed, so did the demand for printed cotton textiles on a vast scale. In England and France, legislation was introduced to restrict imported cotton prints. Yet while Europe attempted to safeguard its lucrative silk-weaving industry, printed cotton from India continued to thrive and dominated the world textile market throughout the seventeenth century (Riello 2010). The European response began with woodblock printing. This labor-intensive process produced complex patterns involving multiple colors, although each color required a separate block. Further developments resulted in copperplate printing which, while it was restricted to a single color, produced additional fine detailing, and the two techniques were generally used together; however, joining pattern repeats was problematic, so lengths of only one repeat were normally produced. Towards the end of the eighteenth century a Scotsman, Thomas Bell, invented the cylinder printing machine that was patented in 1783, and textile printing dramatically changed. Not only was this method cheaper and quicker than woodblock and copperplate printing but also, the repeats lined up, and by the 1820s, with further improvements, sophisticated colors and larger repeats were possible (Tarrant 1987). The twentieth century saw the next important manufacturing landmark with textile flatbed screen-printing and photo-reactive stencils. This involved a gauze-covered frame with dye spread through the mesh by means of a squeegee, and subsequently evolved into a highly successful rotary system in the 1960s. As technology developed, from the mid nineteenth century onwards, so did the chemistry of fabrics, inks and dyes. The combined effect resulted in transfer printing in the 1980s, followed soon after by the first color separations of a digitally produced image being used to create a textile print. The next step was large format inkjet printing, introduced to the market in the 1990s. These flatbed printers gave designers, for the first time, the chance to work with a palette comprising millions of colors, a wide range of substrates from synthetic to natural, an extended range of inks and dyes, and unlimited images and repeats, all of which changed the character of printed textile design.

Chapter 2 explains how textile printing evolved, and analyzes the ongoing technical constraints and key factors associated with printed textiles as they developed to the present day. An argument is made for linking key visual aspects of the patterns produced in a given technique with the materials and knowledge available at that time (Yanagi 1974; Bide 2007); for example, the inkjet technology used for printing an image is the same as that required to manufacture many other specialist products (Hopper 2010). This emphasizes the fact that advanced technology, and its applications, can result in a wide range of different outcomes, depending on the context. Although these choices create diversity, they also lead to a lack of discipline-related focus. As the number of technologies being introduced into textile design and the range of substrates increases, the array of outputs continues to expand. However, partly due to evolving developments in pigment ink production, the quantity of pigment-based textile prints is predicted to significantly rise over the next few years (Quinn 2009). While this may reduce the environmental impact of dye-based printing (Fu 2006), a move towards digital textile printing on this scale will affect the overall character of fabrics from the perspective of both designers and clients.

Chapter 3 details how digitally printed textiles are created, explores the risks to individuality brought about by ongoing digital expansion, and explains why contextual considerations are important when textiles cross over the boundaries of other disciplines. The digital printing of textiles currently encompasses a wide range of processes, from craft-led to technology-based. As a result, artists and designers are increasingly being supplied with new ways of combining and intertwining multiple layers of meaning from both the handmade and the digital. Practitioners have a natural affinity with materials and, while fibers, yarns and fabrics may suggest patterns, structure, form or composition, each material also has its own language (Dewey 1934; Yanagi 1974). Designing and creating digitally printed textiles from a reserve of diverse fibers, dyes, processes and technologies produces outcomes that are informed as much by the practitioner as their materials, and a change in one component inevitably has an impact on all other aspects of the design process (Niedderer 2009a).

Chapter 4 considers the impact multiple materials, such as synthetic and natural fibers, dyes and inks; multiple processes, for example, resist, appliqué or embroidery; and, multiple technologies, for instance, 3D printing, laser printing or inkjet printing, are having on digitally printed textile design. The computer is only a tool with the creative input coming from the artist or designer (Braddock Clarke and O'Mahony 2007). Similarly, studio practice and technology are both products of the imagination but, unlike a machine, artefacts such as printed textiles carry on working imaginatively in the physical world (Dewey 1934). Creative practice requires an initial idea or concept that is developed through the act of forming or making. The kinds of knowledge involved in the studio enquiry can vary depending on what kind of information, or outcome is being sought; however, these may include tacit, procedural or technical knowledge if the practitioner is applying a mastered skill to a new problem or challenge. It may also take the form of experiential or perceptual knowledge, if the aim of this inquiry is to investigate a relationship forged between the artist or designer and a specific artefact. In addition, the knowledge generated can overlap a number of categories, for example when the aim of the investigation is to develop propositional or practical knowledge. This can be achieved through a personal exploration of processes and techniques to convey understanding in visual, verbal and textual formats (Niedderer 2009b). The task facing contemporary digital textile print designers, and practice-led researchers, is to rearrange the boundaries of their studio-based research to more accurately reflect the problems they are attempting to address (Sullivan 2008). With the arrival of the digital era, many restrictions or influences, such as limited technologies, fabrics, dyes and knowledge, are no longer applicable, and the constraints or limitations that had previously shaped the aesthetic of printed textiles, are now largely overcome. Carlisle's pioneering practice-based research into digital textile printing challenges the constraints that previously existed in printed textile design. By investigating a number of new opportunities provided by advanced technology, such as the freedom to create repeating or non-repeating patterns, she is able to break free from the historically binding commercial and technical limitations imposed on the discipline by mass production (Carlisle 2002).

Chapter 5 encourages further practice and research by providing exemplars from cross-disciplinary collaborations and case studies involving different approaches to textile design, for example, generative design (Russell 2013), visual discourse (Treadaway 2012), sustainability (Goldsworthy 2013a), narrative (Rigley 2013), archival (Stephen-Cran 2009), craft and design theory (Carden 2013a; Carden 2013b), and technological advances (Hudson 2014). These positions have been constructed in response to a range of research questions addressing issues of environmental impact, ethical considerations, and the shifting balance between digital and non-digital processes, particularly craft, in contemporary design practice. This chapter also discusses how the act of making empowers the artist with different types of knowledge. These skills inform the creative process, helping to produce new theories, novel opportunities, and provide further research opportunities (Schön 1983).

In Chapter 6 the essence of digital textile printing is investigated. When Heidegger describes the essence of trees, he says that regardless of the species, all trees have a universal treeness (1977). Similarly, if we attempt to define the essence of digital textile printing, whatever the image, type of ink, printer, or base cloth, an inclusive genus also exists. However, as Foucault (1970) points out, certain characteristics of a species are constant while others may occur frequently, but are not always necessary, so in this chapter the constant characters of digital textile printing are identified. The final output of digital textile printing is dependent on numerous factors, and, as a result, its essence must be viewed within the context of all the other materials, processes and machines of its production. A clarification on what is essential for the "digital textile printing" process is provided, along with an exploration of where authenticity lies in the digital printing process (Carden 2011a). This chapter outlines the key components of digital textile printing, including the visual aesthetic of the image, the processes and mechanics of production, the dyes used, and the base cloth upon which an image is printed. It places these material and non-material factors within a cultural context, thereby confirming the authenticity of digital textile printing, and providing a clear framework for future developments in this field.

Chapter 7 explores digital textile printing from a variety of different perspectives. These allow practitioners to develop theories about how and why they create the way they do with advanced technology. It assesses the value of tacit, experiential and procedural knowledge (Nimkulrat 2010; Niedderer 2004a) when used for digital textile printing, and positions the artefact as object so that artists, designers and researchers may extend their investigations into evolving areas of practice and research. It also looks at the implications of the new creative freedom that advanced technology has brought to printed textile design and the subsequent impact on skills.

Chapter 8 investigates how different disciplines engage with similar technologies, processes and materials. Using hands-on craft related practices across disciplines can be seen as a valuable way of making connections with other fields (Press 2007). This allows designers to absorb ideas and expertise from one area of knowledge to another. Studio inquiry is, by its nature, a practical, investigative activity (Barrett 2007a), and is well placed to encourage play, reflection and risk-taking from within the small scale as readily as larger corporate creative environments. By taking a research designer's perspective on digital textile printing, this chapter explores why practitioners in different disciplines, who use the same technology, work the way they do. It analyzes the creative decisions they make in the studio and reflects on recent theories concerning how a range of diverse designers interact with advanced technology (Carden 2015). This chapter also looks for authenticity in digital textile printing and seeks to identify the core processes that are needed to reproduce a printed image. Authenticity is located by drawing comparisons with neighboring disciplines, such as photography and letterpress. In these disciplines the requirement to produce multiple versions of an original print is fundamental, and this feature is one reason why digital opportunities have also led to new ways of experiencing and engaging with practice (Flusser 1983; McCullough 1996; Harris 2005).

This book begins with an overview of how digital textile printing emerged from the earliest days of human abstract mark-making. It considers the tools and materials available to Neanderthals and *Homo sapiens*, and demonstrates how the desire to translate ideas as visible outcomes onto cloth started with processes that continue to be used to this day. Surviving evidence of hand stencil cave art shows the powerful impact that knowing something was made by a single person's hand can have on an audience, and highlights the need for those who use advanced technology in their design practice to be aware of the benefits as well as the anonymizing drawbacks of this new tool. Restrictions that caused many key features and characteristics of traditional textile design, from woodblock printing to rapid prototyping, are described and analyzed. These are introduced to raise awareness of what it means to designers to work in an environment without constraints. The materials, techniques and processes of digital printing are discussed, with connections made between the distinctive qualities of particular fibers, and dyes, and the expectations of contemporary designers and clients. While textile design has ancient roots, digital textile printing sits at an exciting junction in which new avenues are rapidly eliminating older ones and many aspects of its production process are driven by ease, speed and reduced costs. These benefits are often at the expense of taking time to learn, consider and reflect. A series of case studies is also presented that demonstrates the nature of the wide variety of practice that is currently being undertaken by leading designers and artists who embrace advanced technology. These show how practitioners are engaging with research to develop their understanding of the scope and meaning of the creative opportunities digital printing can offer. A number of new techniques are explained that support the value of craft skills, explaining where these can be introduced within the digital textile printing process. The essence of digital printing is explored, shedding light on how knowledge in this field can shape, reveal and link cultural associations. Details are provided of what experienced practitioners feel about advanced technology and how they think it influences their work, what they like about it and what they prefer to avoid. In order to predict which aspects of digital textile printing will survive over time, an exploration into what constitutes an authentic digital print is undertaken. This includes a debate concerning how technology dictates what will and will not be reproducible in the future and how we can create more efficiently when we treat technology as something we can incorporate into studio practice rather than something we feel can intimidate us.

2

HISTORY OF PRINTED TEXTILES

This chapter takes a broad look at the defining points in the evolution of printed textile design. Starting with the discovery of materials that can be used for mark-making and the desire to translate ideas as visual manifestations of emotions, the creative spirit has drawn on the tools and natural resources available. In order to produce artworks social structures began to form, for example, locating sources of pigments and transporting, storing and processing them. Early humans developed different consistencies of pigment-coloring materials and realized they could create a variety of visual effects with them by using their hands or simple tools like sticks; they also learnt to blow through hollow bird bones and were able to spray pigment onto the surface of cave walls. By placing a hand flat against the wall and spraying the pigment solution towards it, they found that a negative hand shape was produced. These hand stencils are to be found great distances apart, indicating that the urge to create in this manner was a natural human condition and not the result of a one-off response to a particular situation. Once fabric construction was developed, from spinning to weaving, base cloth suitable for the application of controlled color evolved. The areas in which fibers such as silk and cotton thrived were early leaders in this field, and the availability of plant and animal dyes and the knowledge to mordant saw these countries heading the advances in textile design. With trade and commerce opening up the potential of global sharing of information and knowledge, as well as the advent of new systems for printing, practitioners were not only able to source new materials and processes but markets allowed them to sell to a far wider audience.

Since the Second World War, artists and designers have embraced developments in computer technology and programing, exploring the patterns and natural order that computers are able to offer (Beddard and Dodds 2009). In weaving and knitting, computers are being connected to machines so that practitioners are able to reproduce finely detailed drawings and photographs in their work (Braddock Clarke and O'Mahony 2007). This provides weave and knit designers with novel tools to create multilayered images and complex fabrics, thereby increasing the range of textures, structures and surfaces they produce. In textile printing, computers have enabled designers to create using an infinite number of colors that can be printed in a single operation, eliminating the need for multiple screens and the specialist skills associated with them. Therefore, while advanced technology has had a major impact on textile design in general, this is particularly evident in the area of digital textile printing and ranges from the creation of images, through production to the fixed, final printed artefacts. Digital processes are redefining key aspects of printed textiles, including the final aesthetic of the outputs, the practice of designers, the expectations of clients and industry, as well as emerging issues surrounding sustainability, craft and slow design.

As culture and technologies developed, the range of materials and processes associated with printing expanded. However, since the mid-1990s, digital textile printing has evolved at an unprecedented rate. The consequences of this rapid progress are changing the manner in which we now create, use, appreciate and understand textiles. Designing for digital textile printing involves many processes and designers are increasingly supplied with novel ways for intertwining complex imagery with unusual base cloths. Digital textile printing is informed by a combination of factors, including the creativity of the designer, the materials and technologies necessary to produce the work; in this context the computer is merely a tool for creating, manipulating and storing images (Braddock Clarke and O'Mahony 2007). Advanced technology represents the practical application of the knowledge needed to form digital prints and, as the technology and prints are both products of the imagination, a series of innate connections exist between them. Textile printing practice is an investigative, hands-on activity and is well placed to encourage play, reflection and risk-taking. However, as with any new technology, the novelty factor must be handled with care until those using it have had time to fully master it. Due in part to ongoing developments in dye and ink productions (Provost 2008), the output of digital textile printing is predicted to rise significantly over the next few years. This shift will affect the character of textiles from the perspective of the designers and clients (Carden 2011b), but it also risks a loss of individuality and therefore challenges the nature and infrastructure of tomorrow's textile industry on a global scale (Quinn 2009). In order to better understand how digital printing has evolved it is first necessary to identify key materials and processes, and understand from where each of them originated and how and why they evolved the way they did.

ORIGINS OF SURFACE DECORATION

Evidence gathered from the numerous examples of cave art discovered in sites across Europe, Namibia and North America (Perez-Seoane 1999), plus the remains of pigment processing workshops dating back 100,000 years in South Africa (Henshilwood et al. 2011), demonstrate that there was a natural motivation to embellish upon surfaces, and that this creative impulse was widely experienced by Palaeolithic peoples (Breuil 1952). As Biggs (2002) suggests, although it may not be possible to understand exactly what they were attempting to communicate, or why, it is nonetheless possible to obtain visual evidence of how these images were produced from evidence that has survived on the walls of caves and stone surfaces at multiple locations (Lawson 2012). The cultural context is not clear, however, as the outcomes of artistic practice are not able to reliably convey this type of information when examined in isolation.

According to Pike et al. (2012), at a microscopic level, any painting, or artefact that has been painted is a suspension of pigment combined with a binding agent such as oil. Johnson (2003) suggests the first colored images could have been simple clay marks on rock. This evolved from scraping and then chiselled gouges on different types of stone, which even at this early stage would have provided them with a variety of mark-making options. Once color was added, the Palaeolithic cave artists of 15,000 to 40,800 years ago were able to use a range of techniques, including engraving, smearing, painting and spraying for applying their images onto surfaces (Clottes 2003a).

Until recently early cave art could only be dated if it contained organic pigment or binders, because these materials are necessary for carbon-14 dating, radiocarbon dating. However, images, now including etchings that contain no organic matter, discovered on stone walls, can for the first time be dated by measuring the uranium isotopes within the tiny calcite flowstone deposits, similar to stalactites, that have formed during the intervening years as a coating or layer over the etched or painted surfaces (Pike et al. 2012).

The oldest cave art found so far in Europe is the El Castillo cave in northern Spain, which is estimated to be 40,800 years old; it displays multiple stencil negative hand images that were spray painted in red and black; these were created up to 10,000 years before those in the Chauvet cave, located in the Ardèche region of southern France that was, until recently, believed to be the oldest in Europe at 35,000 years (Clottes 2003b). Prior to the new uranium-series dating techniques being used during the project led by Pike et al., and funded by the National Environment Research Council (NERC), the Chauvet cave had been named as the Birthplace of Art (Clottes 2003a).

The Palaeolithic European cave paintings of 16,000 years ago, at the Ambrosio cave in Altamira, northern Spain, are judged to be particularly skilled. Although the earliest paintings of the Ambrosio site date back to 34,000 years, it is not until the more recent bison artwork there of 16,000 years ago that techniques are judged to have culminated in high levels of mastery; this leads us to conclude that from 40,800 to 16,000 years ago, that is for over 24,000 years, those people were developing and refining their skills with pigment and mark-making processes and techniques that we are still able to see on the cave walls today (Henshilwood et al. 2011).

While the pigments and tools used are clear, from the remaining evidence on the cave walls, the positioning on the walls indicates that the activity must have been a highly organized social activity, requiring platforms, assistants and large quantities of valuable oils and fats for lighting the areas while the artists worked (Johnson 2003). This meant that maintaining such a culture required information to be relayed from generation to generation and across social groups (Somel et al. 2013). Research by Jezequal (2011) has ascertained that there were two distinct types of pigment mixtures used: a kind of chunky crayon made of pure ochre, which was applied as an outline for the images, while the pigment that was combined with a binding agent was painted inside to fill these shapes. Further study has revealed that at the French caves, these artists may have applied a pre-painting coating onto the surface of the cave wall before painting on top of it, to help prepare it for painting.

It has been speculated that the earliest methods for applying a pigment-coloring solution to a surface were to spray, to smear, to dab or to brush (Clottes 2010; Pike et al. 2012). This implies that spraying was the first non-contact process for the application of a coloring material onto a surface. Speculation by historians is that substantial areas would have been too laborious to perform by this method and therefore handfuls of a plant matter such as moss could have been used to form a hand-held cluster for dabbing the color on, or possibly even by grouping the fingers together and applying via cupped hands (Johnson 2003). If finger-tips or moss were dragged across the surface it would have created a blurred effect, otherwise it could have been applied in single, discrete contact actions, as a kind of early print similar to a rag-rolling motion. The fine bird bones that were discovered in a number of prehistoric sites indicate a skilled use of blown distributed pigment marking reminiscent of contemporary airbrushing techniques. Also, the images of negative hand stencil shapes in the French cave of Roucadour are between 24,000 and 28,000 years old, and were painted in black from carbon and red derived from ochre (Ospitali et al. 2006); while the Spanish cave of El Castillo is even older, at 40,800 years and contains, therefore, the oldest negative hand stencil spray paintings ever discovered (Pike et al. 2012).

The prehistoric pigment of the Roucadour cave was recently examined using Raman microscopy and the color found around the negative hand shapes was revealed as ochre combined with oil or wood, probably in the form of charcoal (Ospitali et al. 2006) There was a mixture of materials naturally formed in the vicinity such as ochre, plus processed ones like soot (Roebroeks 2012). The color range would have started with red, black, white and brown, from iron and manganese oxides and mica, with the pigment ground between stones and mixed with water, saliva, blood, oils and animal fat. There is extensive evidence from caves located in South Africa, Texas, France and northern Spain that prehistoric peoples traveled great distances to source the pigment for their cave art; in the Lascaux region, this was up to 25 miles (Tedesco 2000).

Testing by scanning electron microscopy, X-ray diffraction and proton-induced X-ray emission has discovered that the recipes used by Palaeolithic artists were complex but were consistent across sites; they comprised a pigment for color, some type of mineral extender to make the mixture stick to the surface, and a binder to help provide the appropriate consistency (Lawson 2012). In the El Castillo cave of northern Spain, the cave art has been shown to have been undertaken in four distinct phases: first the sprayed negative hand stencils in red. These predate the second phase of large red outline drawings of agile deer-type animals; then, third, red outline sketches of other animals; and, finally, polychrome images of bison created by multilayering black outline drawings over existing red painted versions of these animals (Breuil 1952). In caves such as the Cosquer, in southern France, two paintings of bison on the same wall were discovered through carbon dating to have been created 8,000 years apart (Lawson 2012). What this shows is that these cave paintings were carried out and skills developed over extremely long periods of time, involving vast numbers of generations, and were both revisited and reworked, starting with the red negative hand stencils, the two large red outline images of animal shapes, followed by multiple small red animal paintings, and finally red painted bison overworked with black outlines. This also establishes the spraying technique of the negative hand stencil images as predating the drawn and painted subsequent images, so the spray-painting was the earliest of the painting techniques so far discovered, and it possibly began with finger fluting before bird bones were used to produce images of greater clarity (Kleiner 2011).

As a contemporary designer, one might ask why Palaeolithic cave artists, whom it appears widely placed their hands on the surface of the walls and sprayed over them to create the shape of the outline as a negative, thereby covering and coloring their own hands in a pigment solution, did not routinely turn their hands over and place, impress or print that pigment covered hand onto the wall more often? Perhaps the binding agent only worked sufficiently for it to adhere for one application only. Transferring the pigment from the hand to the wall would make it a second application, possibly like an adhesive label that sticks on the first application but does not bond so readily when removed and replaced on another surface. When woodblock printing was superseded by copperplate, one of the main problems for textile printers was the different viscosity of the coloring agents that were needed (Cooper-Hewitt 1987). Blowing across a pigment solution that is cupped in the hand and directing it towards a wall produces shapes with little definition; however, by using the hand as a stencil early people were able to achieve a far more distinct, clear shape. This was proportionally accurate, because it represented an actual hand, had a well-defined outline, could display different fanning arrangements of the fingers, and could be replicated with ease to form extended patterns and widespread coverage of the negative hand images. Also, while the bison drawings would have required a high level of intelligence and advanced technical aptitude to create the image, retention and representation and realization of the animal outside the cave, the hand was already there, so its shape and availability were not issues that needed prior working out.

The pigments used to decorate and paint on the surface of the caves were also an indicator of the complex conceptual awareness that marked a significant step in the evolution of human cognitive processes (Henshilwood et al. 2011). Sourcing, storing and combining substances required sophisticated social skills and in 2008, a workshop dating back 100,000 years was excavated in South Africa in which red and yellow ochre had been processed into a paste and stored in shell containers. Also discovered were bones, charcoal and tools for grinding.

This toolkit was similar to the discoveries from sites spread across the globe, from Texas and Tennessee (Dutton 1992; Simek et al. 2012)), France and Spain (Breuil 1952; Clottes 2010; Pike et al. 2012), and East and South Africa (Henshilwood et al. 2011). More surprising perhaps is the fact that the South African cave at Blombos, where a 100,000-year-old ochre workshop was discovered, did not contain any wall paintings. Scientists have speculated that this particular type of pigment manufacturing may well have been intended as an appreciation of color. Similar red ochre has been found in Mousterian graves (Perez-Seoane 1999), and for decorating the human skin (Jablonski 2004). The dating of this workshop also corresponds to the period when it is known that humans were starting to consider ways of expressing emotions and ideas (Lawson 2012).

EARLY FORMS OF SKIN DECORATION

The use of pigment to color and decorate the human body evolved in different ways in a variety of cultures (DeMello 2000). In North and South America color was pricked as individual dots; in California it was scored as a type of drawing, literally drawn across the surface of the skin; in Siberia the pigment was coated onto thread that was stitched and pulled to create colored dots; the Maoris used a cutting, scoring action to mark the skin, while in Polynesia the colorant was tapped in an arrangement similar to that of a rake to form patterns (Jablonski 2004).

This meant that it was possible to produce designs comprising colored individual dots, scored drawings and pre-designed patterns; the dots created units, scoring produced mark-making, by sideways movements. Applying pre-designed or previously worked patterns formed on a secondary material and used as a stamp is particularly interesting from a printing perspective, because the group of dots on the rake-tool formed a motif within the overall pattern (Jablonski 2004), and a similar version of this small motif can be seen in the European copperplate monochrome prints of the eighteenth century.

Traces of early tattooing and body marking have been found in many locations, for example in Japan, Siberia, Egypt, Europe, South and North America, as well as in the Pacific. Using colored pigments from the surrounding areas meant these people were able to create artworks with different types of meaning. Through body art, applying color to themselves rather than the walls of the caves, they could display a variety of images on their own bodies as a mobile canvas. The tattoos were permanent, so not only would these persons have to live with the image for the rest of their lives, but so too was the message their individual body art conveyed to the rest of the social group. The decision of what image to have tattooed had implications for both the recipient and the audience in general. The skin, says Jablonski (2004), is an extremely large surface that protects the body and is the interface between it and the environment surrounding it. So, developing patterns and motifs on this relatively big area in a manner that reflected the materials and tools available to these early humans, within a cultural context, helps to explain why surface pattern began to evolve the way it did.

Few examples of tattooed skin survive, although illustrations on clay fragments provide much information regarding location on the body, as well as geographical location, type of markings, and the shapes and patterns created. A number of notable mummies preserved in permafrost also supply first-hand evidence of the widespread practice and complex imagery involved (Tedesco 2000). For example, a frozen tattooed human man was discovered in Austria dated at 5,300 BC, while in Siberia the 2,400-year-old tomb of a heavily tattooed mummy was unearthed in 1947. The latter find, known as the Scythian Chieftain, is highly significant due to the elaborate well-preserved tattoos covering a large percentage of the body, but amongst them was a distinctive shape of a ram on his right arm. Also in the tomb was a collection of items, including a felt cut-out shape of a ram that exactly matched the size and shape of the image of the ram on his arm. The felt ram had been soaked in colorant and placed onto the skin as a stamp (Hermitage 2013).

Although early cave artists had occasionally coated their hands in a pigment solution and pressed them onto the surface of the walls to leave an impression, the majority were negative stencil shapes. The main significance of this felt ram, however, is that it was crafted as a distinctive image from a material that was itself a manufactured fabric, that is, the felt was constructed from wool and then cut and the resultant shaped piece used to apply a coloring agent in an exact predetermined form, with pressure, as a single action onto the surface of the skin. Unlike the negative hand stencil shapes on the cave walls, this ram was not a stencil but an actual shape that was crafted and created and then used to transfer a stain in the exact form of a previously worked animal shape. This mummy required a series of complex processes to decorate it: felt making, cutting and shaping, pigment preparation, image drawing, printing and tattooing. All of these individual processes required significant levels of skill, as did the sourcing and production of coloring materials. Technical proficiency with a range of tools was required for both stonewalls and soft, porous human skin. Knowledge of felt construction was needed, which in turn required a sufficiently high quantity of domesticated animals to be at hand in order to produce enough fleeces for the society to master the skill of felting. Also, a nomadic lifestyle was required to allow the tribe to travel with their herds in search of grazing grounds, which in turn led to the people of the group being dependent on felt for clothing and movable tents for survival in the first place. Therefore, the red dyed ram image on the chieftain's arm, and the felt cut-out, contained a great deal of tacit, explicit and procedural knowledge.

The coloring agent on the cave walls was pigment-based; however, possibly through picking berries and collecting beetles, early humans began to become aware of the coloring potential of organic matter in the form of dyes. Pigments are made up of largely inorganic substances that can be ground and mixed with liquids, such as water, oil, fats, egg yolk or honey, but do not completely dissolve; dyes on the other hand, are derived from organic matter – for example, plant juice and crushed insects and minerals which are soluble (Roy 1978). The range of colors from dyes included red from madder, yellow from turmeric and saffron, blue from indigo, purple from kermes and scarlet from cochineal, depending where in the world the prehistoric people were located (Ye 2000). The nature of the dyes, such as acid suitable for animal fibers, or direct for use with plant sourced fibers, also dictated how and where color could be used in textile design.

DECORATION ON FABRICS

In Ancient Egypt four main processes were developed during the Dynastic period: smearing, by which the coloring paste was spread across the surface of the fabric and pressed into the fibers; indigo dyeing, a complicated vat process involving oxidation; adjective dyeing, in which a mordant is added during the coloring process to fix the color to the fibers, probably alum which is a readily available salt in Egypt; and, dyeing twice, whereby a fabric is dyed one color and then dyed again with a second, resulting in a combination of the two. These people also understood how to remove color, possibly by leaving fabrics in the sun (Vogelsang-Eastwood 2000).

Samples from the tombs show little evidence of fibers being dyed before fabrics were constructed. Usually linen was either spun and dyed, in which case unrolling the thread reveals un-dyed fibers in the center, or else the final cloth would be dyed after weaving, and for this a movement of the threads, particularly in denser areas of the cloth reveal less dark or un-dyed areas under the cross sections where warp and weft overlap.

Dyeing techniques developed in different regions depending on the source materials for adding color and also the types of fibers available. In India, for example, a strong red was produced that required twenty separate processes and was called Turkey Red (Tarrant 1987; Natural History Museum 2013), while a vibrant green was created from copper steeped in fermenting grapes. The use of mordants like tin, alum, chrome and iron extended the range of colors that could be obtained on fabrics. The three main techniques for applying mordants during dyeing are:

1 mordant the fibers or yarn and then dye in a dye bath,
2 mordant and dye together in the same container, or
3 dye the fibers, or yarn, and then mordant afterwards, to fix the color.

Natural dyes are either adjective dyes meaning they need to be used in conjunction with a mordant, in one of the three systems above, or they are substantive dyes, in which case they do not require a mordant to fix the color, as they are adequate on their own (Dixon 1979).

WOODBLOCK PRINTING

Benjamin (1936) points out that the act of stamping or printing is significant because it may be used for an individual creative one-off output, but most importantly, it also enables the means of reproducing a print to be repeated multiple times. He also explains that the material that was most useful for printing, and therefore key to its development, is wood. By the Edo Period, explains Yoshida (1939), the artist who created an image, the cutter and the person who printed it, were all acting in very distinct ways. In manufacturing terms, this speeded up production, but from an aesthetic perspective it was problematic, because while traditionally an artist would control every part of the process, from the idea of the image through to the final print by themselves, fragmenting the system so that each aspect was performed by a different specialist meant that there was now a lack of continuity in the creative process. The artist who had the initial idea in mind would produce a sketch. The cutter would then make a version in wood that was a personal interpretation of that sketch. Finally, the printer would apply color and print each block to his own vision of how the final print should look. This meant three stages and three different specialists existed and separated the original idea from the final outcome. Also, the artist had to think, or anticipate, all the steps that the cutter and the printer might have to make. For example, if an artist was not familiar with the type of wood being used, he or she would be ill prepared for the thickness and quality of the line that could be expected or achieved, or the depth of cuts or the wood's natural characteristics. Also, if the wood had been sourced from the center of the tree it would be of a different density to that cut from an outer section. The artist would first draw the sketch on paper with a fine black outline and indicate with an annotated visual key how, where and in what colors, they desired each area to be completed. However, by this time, the process as a whole was fractured with ownership difficult to attribute, so the outcome was a combination of characteristics reflecting the three specialists involved.

Like weaving, printing is believed to have evolved in Ancient China. It was initially developed to utilize woodblock techniques for the reproduction of first Buddhist images and later images plus texts. In China, by the ninth century, the printing of books had taken off to such an extent that the large quantity of surviving printed material meant they were common everywhere in the country. Interestingly, while there were width constraints, scrolls were unlimited in terms of the length that could be produced; but this was to be replaced by a uniform vertical and horizontal format, similar to that of the computer monitor, and resulted in a book layout we follow today. Although printed books were widespread in China, long before Gutenberg, the pictograms used in texts for communicated visual language were highly complex; if Chinese pictograms had been digital, that is, divisible into interchangeable characters, similar to the alphabet of the West, then printing with movable type might have taken off earlier in China rather than through Gutenberg. Printing provided the opportunity to produce multiple reproductions of books and also to print money. Both of these impacted significantly on society and cultures around the world, but it also marked a major turning point for the printed textile trade that grew up between India and China in the East, and Europe and America in the West.

Woodblock printing on textiles was practiced in Egypt over 2,000 years ago (Chamberlain 1978). The woodblocks used for textiles were, at first, constructed of simple designs and made use of the relief effect created by carving into the surface of the wood; printing on fabrics was established long before paper was invented in China around AD 105. By the fourth century AD techniques were well advanced with evidence existing that demonstrates this was also the case as far afield as India, China and Japan in the East, across to Mexico and Peru in the west (Rehbein 2010). Instead of ink spread on shaped wood and then stamped down onto the substrate, with paper the ink was applied to the wood as a paste and the paper laid out flat on top of it; this was then rubbed to transfer the inked-up image onto the paper above, rather than the other way around; in this way the handling of the substrate controlled the transfer process rather than the stamp. Further developments in paper manufacturing techniques by the Spanish led to a rapid rise in printing and this shifted the emphasis of production towards Europe, in particular Italy, Germany and France; and, as a consequence, the spread of reproduced pictures meant that imagery from traditional sources was now communicated widely.

While the skilled woodblock cutter acted as an interpreter of the artist's original drawing, over time the artist created work that was more suitable for woodcuts as they had the knowledge and skill to understand what worked best for that medium. As demand increased, printing in multiple colors grew with a block for each color, at first two colors, usually pink and green. The wood selected for intricate cutting was sourced from the hardest central area of the tree, and was therefore limited in size. The patterns that were produced in this way comprised small repeats. The outer parts, being softer, were chosen for the less finely detailed color sections. Another issue was that the tools of the artist were very different from those of the cutter, so the translation of image from sketch to cut wood could also vary a great deal. Woodblock was often used with additional hand painting in order to add small areas of color (Sardar 2000). Woodcarving was widely used, although one of its main markets was the printed textile trade with copies of elaborate, exotic hand-painted textiles from the East and finely woven patterns produced in France during the eighteenth century (Robinson 1969).

The wooden blocks used for textile printing came in different shapes and sizes, although the most popular were rectangular measuring approximately 25 centimeters by 30 centimeters, roughly the proportions of today's computer monitors. The blocks could also be modified by introducing strips of copper or brass and even included felting so that larger areas could be colored. The felt was packed into the cut-out regions of the wood and when saturated in dye would give an even covering of color to the print; the combination of carved wood, metal strips and felt packing provided a useful range of textures for the designers to work with. The final prints also displayed differences of color uptake on the back and around the lines of the felted impressions (Cooper-Hewitt 1987). The printers would use techniques to disguise the registration dots needed to align repeats, and motifs such as seed heads were incorporated to make these less obvious to the client; also, in order to train the eye away from the joining of repeats on the vertical, designs were created with off-set patterns or incorporating dominant images of foliage to disguise the units that were being repeated. Different countries began to develop their own individual styles and imagery, such as the English floral patterns or the French aim to create textiles that looked as if they were Indian printed cottons.

CHINTZ PRODUCTION

Trade along the Silk Route brought not only Indian printed textiles, but also sheets of patterns and designs for the Western market to imitate. The European producers of the eighteenth and nineteenth centuries were copying the Indiennes or chintz designs from the East that become extremely fashionable. However, chintz production comprises highly complex patterns, traditionally produced using woodblocks, resist dyeing techniques and hand painting to achieve complicated color combinations and images. No matter how closely the textiles were copied, and the motifs and colors convincingly and accurately reproduced, the versions manufactured in the West tended to fade or run easily and were not fast to light or water. When traders from the French colonies began to bring back details and instructions from the east coast of India regarding indigo vat dyeing, in which oxidation is needed to produce color, and white resist dyeing, they were now able to create washable, colorfast versions of the prints of their own to challenge the original chintz textiles from India (Sardar 2000).

Knowledge about dyes and mordanting that was widely available in India influenced the colors and images that could now be produced. As Prakesh (2012) states, for centuries the primary aesthetic in Indonesia was deeply rooted in textiles; their trade, methods of production, techniques, materials, design and motifs. Ikat in particular, a process whereby the warp and, or, weft yarn is tied then dyed before weaving has developed in India and South East Asia in line with local traditions and has therefore become encoded with associations and narratives. This differs from the tie-dye and tie-bleach of other Indian areas in which the woven fabric is tied and dyed, because for ikat the threads are tied and dyed beforehand. Another variation is in Sumatra where the woven fabric is sewn as well as being tied before multiple dyeing to create highly complex geometric and multi-colored patterns (Prakesh 2012).

The precision of the tying or stitching to resist the dye in the dye bath includes a system that pins out pre-designed images onto wet fabric. Each of these processes adds a further stage, increasing the distance between the idea and the final textile. Resisting dye is also achieved through the application of a substance that is impervious to dyeing, such as beeswax, but one that can be controlled and temporarily adheres to the fabric; the substance is then removed after dyeing, and can be repeated multiple times to build up complicated designs and images. In India this was a key process in the complex production of chintz, and it was one of many individual stages that were distributed to different families to undertake. The initial image was drawn up by an artist and then passed to another craftsperson to transfer the design onto the cotton fabric using fine charcoal dust and paper pricked out with the outlines of the pattern in a manner similar to stencilling. The charcoal shapes were used as outlines drawn with hot wax and the fabric was then dyed at a separate location. Originally chintz designs were a combination of resist-dyed and hand-painted processes, but later on woodblock printing was also introduced. The whole multifaceted process also involved a variety of mordanting techniques to fix the various colors being used in each design (V&A 2013). This distribution of processes and techniques has been a characteristic of textile production in many locations since the Early Minoan period. Documents in the form of clay tablets from the Aegean around 2000 BC illustrate the elaborate extent to which the raw materials had reached (Cline 2012). Initially this involved records of fiber production, such as wool and flax, then included types of cloth created and the quantities of raw material and equipment needed to weave them; and, as Herodotus (1998) noted, textile production was a highly organized, specialist craft activity even before the extra complexity of surface decoration evolved to add to it.

COPPERPLATE PRINTING

While woodblock printing began with textiles, and then crossed-over to paper with the development of letterpress, copperplate printing had been around since the fifteenth century for printing on paper; but it was not until 1752, when it was introduced in Ireland for calico printing, that it became extremely popular for textile printing. One of the first problems, compared to woodblock printing, was that the ink needed to be of a very different viscosity to that required on wood. The woodblock comprises a relief surface, while copperplate is etched and therefore comprises scores incised into the metal where the colorant lies. A thicker paste is needed for the wood, so that it does not pour off, and was generally produced using a substance such as starch, while the consistency required to fill in the fine lines of the plate had to be far more fluid. Also, while the wood was stamped down onto the fabric base, the engraved plate was laid out flat and the substrate placed gently on top, and the two materials were then pressed tightly together with a mangle to transfer the liquid ink from the metal plate to the fabric (Chamberlain 1978; Cooper-Hewitt 1987).

The scale of the repeats for copperplate were different from woodblocks; whereas the size of the wood repeat was relatively small, usually around 25 centimeters, metal plates could be created from anywhere between 84 to 104 centimeters up to 2 meters. Clearly a textile with this scale of repeat would be more suited to large interiors than for garment designs, and when the first copperplate cotton printed textiles emerged it was difficult to spot the individual repeated designs as the joins were almost flawless. Engraved copperplate allowed finer detailing to be achieved but was restricted to only one color, so the new technique was often used in conjunction with woodblock printing and hand printing to add further colors to the final textiles. The printed textiles characteristic of this period comprised small repeats of monochrome striped motifs, resembling woven patterns from Turkey, with dots forming a ground pattern. These tiny dots were designed to resemble flowers that had been painted with brush marks. Large quantities of printed cotton using this technique were produced during the nineteenth century, but it was the introduction of mordanting processes into the West that was to make an even more significant impact on the printed textile industry from the first part of the nineteenth century.

The crossover from carving on wood to metal, for interpreting artists' sketches, was as important to the development of letterpress as it was to textile printing. The early type cutters, before Gutenberg, were imitators of manuscripts. Similar to the woodblock copiers of Indian textile designs, their aims were to reproduce manuscripts as quickly and cheaply as possible, while also keeping as close to the aesthetic of the original as their skills and materials would allow (Updike 1962). Interpreting all the different letterforms that the human hand can create was not seen as being particularly important until the end of the seventeenth century, when Jaugeon, in Paris, outlined a series of definitions and guidelines for designing type: initially in a grid of 2,304 tiny squares. The theory was that a single model of every letter of the alphabet would be available so that unskilled woodcarvers could make better, more useable blocks. What it did do, though, was to install a sense of uniformity in the design of type. Also, it was discovered that printing type at the size for which it was initially designed was far more successful than increasing or decreasing the scale afterwards, as each size adheres to its own laws and principles. So on one hand the type from cutters with less skill was improved, but the outcomes from other designers was not always better. The person who drew up the original design, believes Updike, is the one best placed to hand cut the punch of the type and, for maximum effect, it should be done as close as possible to the actual size. Even an individual letter has its stem and hair-line at the outer edges where the ink has to be the desired consistency to stick to the face, transfer to the paper, not squeeze out and distort the image, not be too dry as that would mean not enough to print correctly and not too much or it will run off the face before printing (Wakeling 2013); also, too wet and the process will end up warping the paper and bleeding after printing. For textile printing, woodblock images have similar issues, and each color used needs to fit in adequately and register correctly. As well as designing the type, a skilled compositor and skilled artist are required; and, like textile printing, the substrate or base cloth imparts its distinctive character on the final outcome.

Paper of the eighteenth century largely consisted of fiber. It was handmade from rags, old pieces of cloth that were pulped down and formed on a mesh and pressed to create a smooth surface: so paper of that time could be termed a non-woven fabric. A dandy roller produced indented lines, vertically and horizontally, that gave the paper a woven appearance (Herbert 1980). Gradually paper was made from wood pulp, although rag-based paper was still available, expensive and more difficult to cleanly or evenly print upon.

ROLLER PRINTING

Woodblock and engraved copperplate printing continued its domination of the European textile industry throughout the eighteenth century. However, when Thomas Bell developed and patented his roller printing process in 1783, the popularity soon decreased. Roller printing produces a continuous system that picks up dye from a trough while it prints in a cyclical manner, meaning there is no requirement to print individual blocks or plates and the fabric base is fed through continuously; there is no need to re-register each repeat. The size of each repeat at the beginning was quite small, as it matched the circumference of the cylinder; but this increased over time. Also, at the start the prints were monochrome, just as copperplate had been, and the early prints were designed to mimic the copperplate patterns as a result. However, as the copperplate versions were so much larger, it was relatively easy to tell them apart. As the process evolved, alternative ways were developed to produce large areas of color that allowed the potential of the metal etching to be translated into the images. These included fine shading and intricate patterns that resembled net and were used to produce overall ground detail.

Roller printing was suitable for use with many more types of fabrics, for example wool and linen, as well as cotton, and these were especially popular with the new synthetic dyes that were chemically produced after Perkin's discovery of the first completely synthesized colorants in 1856 (North 1969). These created brightly colored prints in a vast range of hues, although the early chemical dyes faded quickly and were patchy in appearance. Morris's early chintz designs were hand-block printed using these synthetic dyes by Thomas Clarkson and Company; but Morris felt strongly that the images did not always suit the synthetic dyes, and he thought the more limited number of colors, and the increased skill required to handle them, would result in more thoughtful designs than the chemical dyes were producing (Parry 2009). Natural dyes, believed Morris, were better able to harmonize with each other than the aniline alternatives. However, the rapid rise in synthetic dye availability began to undermine the centuries it had taken practitioners to master natural dye and pigment use for coloring fabrics. As a result, the skill and knowledge that had been developed and acquired in natural dye production was still far superior to that of the fledgling synthetic dyes, even though its use was on the decline.

LITHOGRAPHY

In 1798 a playwright called Alois Senefelder from Munich discovered that it was possible to replicate his manuscripts on paper if he first wrote on limestone with a greasy substance similar to crayon. The smooth, porous nature of the stone meant that the waxy script adhered to the surface but when he washed the stone with water, some of this remained soaked into the areas around the text. When he coated the stone with ink, which was greasy and repelled water, the ink did not stick to the moistened stone but did adhere to the script. It was then possible for him to lay a sheet of paper over the stone and transfer the inked script to the paper. Because there was no scraping involved, the process could be repeated many, many times without degrading the quality of the original crayon outlines. The limestone Senefelder used was suitable as it had a calcareous nature that was polished yet absorbent (Bankes 1976). The process was further developed for reproducing drawings, at first in monochrome and then in multiple colors. As Bankes noted, the quality of etchings and engravings, the work of skilled silversmiths or carvers, was more sophisticated than the translations achieved through lithography, and he felt that copies produced using this process never managed to achieve the same spirit as the original drawings. Although lithography was mainly practiced for printing on paper, it was effective when used on cotton and silk fabrics, particularly for smaller items rather than lengths of cloth (Weber 1966). However, creating textiles of multiple colors necessitated separate stones for each color and caused problems due to registration issues.

SCREEN-PRINTING

By the 1930s flatbed screen-printing with gauze mesh stretched over frames enabled designers to print out blocks of solid color. Instead of cutting wood or engraving metal plates for the parts of the image that they wanted colored, the screen-printing technique took a negative approach, much like the stencils of early Japanese prints or indeed the very first-hand stencil images created on the surface of the cave walls by Palaeolithic peoples. Printing had, to a certain extent, come full circle. As for woodblock printing, the individual colors were divided into separations, one for each, with registration marks to ensure that all colors were correctly lined up with each other. Initially lacquer was painted onto the mesh of the screens to block, or resist, the color; later this was achieved using photo-reactive stencils. The dye in a paste was spread or smeared, through the fine holes in the gauze onto the fabric with a squeegee to apply even pressure and coverage. By the 1950s screen-printing had further developed into a rotary system with up to twelve-color textile prints being produced. This technique replaced roller printing for most printed textiles by the end of the twentieth century.

LASER CUTTING

In the late 1960s laser cutting or printing first appeared in a number of industries, although initially for cutting and engraving hard woods, acrylics and paper; the process has been successfully incorporated by textile designers to recycle mundane polyester fabrics (Goldsworthy 2009), and bestow new value to waste textiles through intricate embellishments (Quinn 2012) and up-cycling (Earley 2009; Quinn 2011). Laser cutting provides a range of advantages over traditional forms of printing; it is fast, and, is a non-contact process that therefore, unlike woodblock or copperplate, does not deteriorate (Yusoff 2011). However, while laser cutting synthetic fabrics seals the cut edges, making it particularly suitable for fine cutwork without the need for seaming or unattractive fraying, the sides of the cut-out areas can be rough and unpleasant. The technique does allow for hard edged images that contrast well with both color and texture, and enable practitioners to achieve intricate sculpted three-dimensional qualities to their textiles in a way previously unattainable (Heyenga and Ryan 2011).

3D PRINTING

In 1992 Scott Crump applied for a patent for an apparatus that enabled a computer-aided design and computer-aided software system to produce a three-dimensional object formed from a material that is dispensed from a print head, and then built up and solidified by an adhesive resin. The material can include thermoplastic resins, metal or glass, and the individual layers stick together to create a solid structure (Crump 1992). Crump's invention, built on previous systems and technology, was to join together all the various principles and components that are necessary to enable designers to create three-dimensional objects from digital images.

In 2003, Freedom of Creation (Kyttanen 2010) applied for a patent for a textile construction process that uses layering techniques to create chainmail-type structures (Delamore 2004). The rapid prototyping equipment fabricates a three-dimensional version of the computer-generated model in two ways: one is subtractive in which the object is cut out of a solid block of the sculpting material, and the other is an additive process that builds up individual layers of powder that is interspersed with adhesive bonding to generate the shape. The additive system is the version being developed for textiles, however to date the material is not particularly suitable for apparel design, although the ability to produce items that do not require joining, so eliminating seams, has made the idea particularly interesting to designers. Body scanning data is already fast and inexpensive, so a three-dimensional model of the human form can easily be created and then a three-dimensional garment made to fit. However, skin-tight apparel is perhaps most suitable for medical and sportswear ranges. The process does have the option to be crafted from starch powder, so could become more popular as a sustainable option, once the adhesive can be developed with similar principles in mind. As an alternative system for textile construction, laser sintering may one day be developed to solve long-term garment producing issues (Delamore 2010); but Grossman (2013) has concerns about the process being able to convince an audience expecting textiles with traditional tactile qualities; as she explains, manufacturing structures for the lighting market are progressing well, as is laser-etching on glass, but the same is not yet the case for textile design (RapidToday 2013).

DIGITAL PRINTING

The first digital textile printing was developed for carpet manufacturing by Milliken in the 1970s. This process involved blowing air to propel droplets of ink towards the surface of the carpet. Around the same time punched paper tape was being used that evolved from the punch card system originally devised and patented by Jacquard in 1801 for intricate loom weaving. IBM continued to use printed punch paper until as late as 1984 (Tyler 2005), but by then digital printing had begun to demonstrate the key features that attracted widespread interest among textile practitioners, particularly for sampling. These characteristics included thousands of colors, non-contact with a substrate, non-constrained scaling of images, and an unrestricted size of print runs (Keeling 1981). With the development of transfer printing in the 1980s, images of increasing complexity could be printed on paper then, using heat, be transferred onto fabrics. Towards the end of the 1980s, a computer program was written at Dundee University that allowed digital images, created by simple programing, to be output as color separations from a dot matrix printer, using RGB (red, green, blue) channels. In this way it was possible to screen-print the area of the pattern represented by each color separation, and produce a hand printed version of a digitally created image (see Plate 7). This provided designers with the wherewithal to create new coded images, such as finely detailed, complex shapes including geometric progressions that would previously have been far too complicated to separate out by hand into individual colors. The images were displayed as a light source on the monitor and also could be conventionally screen-printed to colors that the designer could both select and control. With the screen displaying the image as a division into pixels, and the image into tiny dots, the digital nature of the relationship between the screen image and the printed images was evolving.

The 1990s saw a rapid rise in the interest of digital inkjet printing, and Stork launched its Fashion Jet at ITMA in 1995 (Moser 2003). From 2000 onwards software and hardware developers continued to work on the remaining challenges that stood between digital printing and conventional screen-printing, such as print speed and color cost.

DIVISION OF LABOR AND IMAGE

Palaeolithic cave art, at first the expression of mark-making by individuals, developed into distributed labor with pigment preparation conducted in a workshop and requiring organization, sourcing and procuring of base materials, grinding, mixing and storing the quantities and consistencies that were required. Once the materials and tools were on site, the artist created the negative hand stencil sprayed images; there was no preparation needed for the shape, the hand was there ready to be sprayed over, and pre-existed in the necessary format for the final image. Over time, however, textile printing evolved to encompass a distributed textile labor force (Marx 1976): woodblock with artist, cutter and printer, in which cutting fine details out of wood was a highly skilled task, resulted in outcomes that reflected the characteristics of the wood and its affinity with dyes and hand printing, but it was also a problematic material as it would ultimately wear down or even break during printing; copperplate with artist, engraver and printer. However, there was difficultly with registration and joining of repeats, so patterns were usually complete in one repeat and monochrome: roller printing with artist, engraver and printer, in which small single color patterns, striped motifs and tiny patterns, one color initially, then up to six 1-inch to 80-inch repeats in two to six colors in a single operation; chintz printing with artist, pincher, dyer, painter, block cutter and printer, producing much sought after designs but involving multiple steps and hard to replicate accurately; screen-printing with artist, lacquer applier and printer, outputting textile prints of up to twelve colors.

As with the fundamental principles of movable type, such as Romaine de Roi, created for Louis XIV in 1692 for use by the Imprimerie Royale, and described in 2,304 tiny squares, the image on a computer screen is also designed to exist within a regimented frame and whatever scale the final print is intended to be printed at, it is only seen on the monitor up to a limited size depending on the dimensions of the screen. So, with digital textile printing, the hardware, software, fabric and dyes are generally prepared by persons who are not the designer and, other than using or selecting from these options, the designers are not personally involved in making them but are nonetheless now able to create with these materials, technologies and processes in mind, and they are able to design the final digital textile prints by themselves knowing that technology, including the form of a large format inkjet printer, will print it out without them even needing to be there, for the first time since the Palaeolithic cave artists sprayed color over their own hands as a stencil on the surface of the wall. The hand stencil used 40,800 years ago on the cave walls is striking for a number of reasons, but from a textile designer's perspective a key feature is the context of the image. Unlike the later geometric and abstract shapes painted on the surface, the hands are the nearest to photographic representations or realism. Also, creating artwork with a movable, ready formed shape such as the hand, one that requires no further carving or working, and is already in the environment, is similar to today's preoccupation with taking images from the virtual landscape and being able for the first time to print them without having to first rework or modify them. Like digital textile printing, it involves spray-painting rather than contact transfer printing, and as a result, in common with inkjet printing, the image needs to be fully worked out before the act of spraying the colorant takes place. Therefore, in terms of their shared characteristics, I would suggest that digital textile printing is more closely aligned to the negative hand painting on cave walls, its most distant relative, than it is to its closest, conventional screen-printing, the one it was developed to mimic and supersede.

CONCLUSIONS OF HISTORY OF PRINTED TEXTILES

This chapter has shown that the major developments since cave painting in dyes, materials and techniques have been evolving amid the desire to communicate, to visualize and to utilize tools to make artefacts and convey meaning in a cultural context. These advances have evolved at different rates within a variety of disciplines with achievements in one impacting on and promoting further developments in another. Against a backdrop of discoveries in all its nearest disciplines, textile printing has been well placed to take full advantage of progress in dye chemistry, fabric production and digital technology.

3
TECHNOLOGIES, SUBSTRATES AND DYES

In this chapter the range of materials and technologies needed to produce digital textile prints is explored. Many textile processes are explained in order to identify where different tools and dye knowledge have come together to provide us with extraordinary products, such as the isolated regions on the east coast of Indonesia where a unique form of silk batik has evolved, or the drive to develop synthetic dyes in order to satisfy the growing demand for printed textiles, or the multiple processes and dyeing techniques undertaken in the production of chintz. The skills necessary to design with natural and synthetic, non-digital and digital, stem from these basic principles and if we are to create effectively with the materials available to us, then there needs to be a better understanding of what knowledge has already been realized in this field.

The inkjet technology currently being used for printing images onto fabrics is the same as that required in engineering design to manufacture circuit boards (Hopper 2010). Technology is developed by humans to fulfil specific functions, claim Polanyi (1974), Heidegger (1977) and Pye (1978), and it therefore produces outcomes that are predetermined. When advanced technology is used for digital textile printing, the results are often described as flat, lacking surface interest (Brandeis 2003). Pye (1978) refers to this uniformity as the workmanship of certainty; which can help to explain why the final digital prints are sometimes not as successful or interesting as practitioners or the client might expect. Heidegger (1977) acknowledges the human influence over outputs of technology, while at the same time he and Yanagi (1974) maintain that although machines can be relied upon to produce consistent results, they are not able to replicate the human hand. Phillipson (1995) also observes that for practitioners, the challenge is to free themselves from the constraints of what is implicitly expected of them. When a task is taken out of our hands, technology is increasingly seen as an instrument; as Eco (1989) explains, once human beings create machines they immediately feel oppressed by them and the non-human relationship we have with technology is so unpalatable that we instinctively attempt to merge beautiful features with the functional aspects of machines in order to forget we are governed by them. A large format inkjet printer is not designed to adorn a domestic space, and while it has a usefulness that outweighs its visual appeal, the fact that we treat the printer as a non-human machine implies we are likely to feel enslaved by it (Heidegger 1977; Eco 1989). Until designers are comfortable entering into a dialogue that raises the current relationship above that of the purely functional, they are unlikely to produce textiles of the maximum possible creativity.

McCullough (1996) suggests that when technology is used in partnership with human input the results are more interesting than those achieved from technology alone. A constant dilemma for digital textile print designers, in common with disciplines such as engineering, is that a designer must be able to respond quickly to ever changing circumstances that are occurring in advanced technology, while simultaneously being mindful to address the evolving needs of the society for whose benefit they are working (Brooks 1967). While such machinery requires those charged with operating it to be provided with clear instructions and directions so that they may fully under-stand how best to use it and appreciate its advantages (Turing 1950), advanced technology does not evolve in isolation and the cumulative effect of merging existing systems with other disciplines increasingly creates new opportunities.

As practitioners, technology is most useful when we embrace its full potential in ways that work in conjunction with our own practice (Polanyi 1974; Dormer 1997). However, it is necessary to identify which elements combine to make a digital print and where cross-disciplinary technologies currently exist in order to maintain a robust base for the domain and suggest where, and how, future techniques may be deployed. Turing (1950) says that a computer does not make mistakes, therefore it is incapable of producing something it was not expected to do; although, it can be made to produce something that appears to look like a mistake, that is not really an error but is a planned effect that looks like an error (Marcus 2008). This is what practitioners are suspicious of even though they also understand that experts know what is true, and what is not. The things a designer can do that the digital process cannot include acting instinctively and reflecting or perceiving, in other words, things not explained by rules (Turing 1950). This is why a computer cannot convey skill or connoisseurship, and an inkjet printer that produces a digital textile print does not convey any emotion or feedback about how the artefact has turned out, meaning that from the machine's perspective, nothing has been learned from the creation of the digital textile print (Polanyi 1974).

Compared to a computer monitor, the digital textile print has permanence and physical presence, characteristics that Tufte (Zachry and Thralls 2004) prefers in the form of paper over a computer screen. He points out that the monitor is overrun with multiple space-taking interface administrative clutter, and that the software imposes a cognitive regime on those who use it; this is further compounded by the fact that, as Mirzoeff (1999) notes, our environment is actually onscreen nowadays, so that is where our emotions and experiences are most likely to be focused. According to Gale and Kaur (2002), constructed fabrics, such as woven textiles, took thousands of years to evolve. They are believed to have developed from rope making, a process that first required manual dexterity, twisting, and the forming of fibers into yarn. Recent analysis of a fossilized hand bone found in Kenya shows the ability of early humans to manipulate using their hands in this manner was around 1.42 million years ago (Ward et al. 2013). The structure of weaving, warp threads stretched vertically interwoven with weft thread crossing horizontally creates a firmly held together web. The structure of the weave generates idiosyncratic imagery of abstract, geometric shapes following the vertical and horizontal threads of the weaving (Yanagi 1974). There are physical features and intrinsic characteristics of different fibers that indicate why methods of construction and dyeing processes evolved and developed in key regions of the world. For example, silk has been known for approximately 6,000 years, and the silk filament is strong, very fine, and can be unravelled in a single continuous thread of 1,000 meters (Fenner 1991). Silk fiber dyes readily and so it is not surprising that this thread was instrumental in the development of weaving in China from as early as 4000 BC.

In India cotton has been used for fabric construction since between 5500 and 4300 BC, even though its cultivation and preparation for spinning and weaving is far more complex than silk. The freshly picked cotton requires cleaning, combing and carding to rid the fibers of the debris and impurities of agriculture, and to tease out and untangle the fibers. The cotton is then spun and prepared for weaving into warp and weft threads. Only then is the yarn woven into a cloth, and the entire process involves many laborious steps. Compared to silk manufacturing, cotton would have taken considerably more time to develop and perfect (Prakesh 2012). Unlike silk or cotton manufacturing, which developed in areas where these raw materials were prevalent, wool was not widely used to produce fabric in many of the locations in which domesticated sheep were plentiful. Wild sheep and goats would not have been providing large enough quantities of raw fiber for a well-developed felt industry to emerge (Laufer 1930); for example, felt was unknown in Africa, America and Peru, even though wool was widely available in all of them. The reason, historians suggest, is that felt only evolved in cultures where the characteristics and qualities it possessed were paramount to the survival of the society itself. For example, the nomadic tribes of Central Asia depended on felt for clothing and shelter; felt is highly wind- and waterproof, retains heat and is readily transportable enabling them to easily dismantle their accommodation and follow their flocks and herds to new grazing grounds. Colored felt is cut and placed on top of other already produced felt, and is then beaten or rolled into it, not painted; so once the felt is dyed, both the base and the additional colored piece in its final shape are merged together, and no pigment mixing in the form of a paste or solution for printing or painting is required. As a result, it is no surprise that textile printing did not emerge from the textile practices of the nomads of the Steppes.

COLOR IMAGES

Historically, spraying and smearing needed only basic techniques to combine pigment with binding agents so as to produce colorants of varying viscosities. However stamping and printing required a further development stage involving a pre-formed image-making device designed and constructed from a secondary material. This allowed color to be applied to it and, on contact, stamped the colored image onto the base surface. This process provided the practitioner with time and artistic freedom to experiment and plan beforehand on a physical version of their initial idea prior to applying it to the final surface.

STENCIL IMAGES

Early Japanese printed textiles demonstrate a skilled application of cutting two-dimensional shapes in the form of stencils (Gale and Kaur 2002). The stencil allowed the artist to produce an even covering of color within the outline of a shape, but it also enabled the pieces to be reproduced. However, the image as an idea was similar to the stamp or printed image in that it had first to be worked out on a first material, was formed by cutting often on a second material, and this was then used to create a further translated version of the original image on the surface of yet another material, such as fabric.

RESIST DYEING

In countries like India, the availability of fibers such as cotton and silk, sophisticated weaving techniques, mordanting and dyeing processes, resulted in a series of resist dyeing systems that produced highly complex designs and patterns, for example tie-dyeing and batik.

Tie-dye

By first tying a plain fabric using thread and then dyeing it, when the cloth is unwound, the process creates a decorative pattern. Depending on the complexity of the tying, many elaborate images can be created from this resist method of dyeing. Different folding and tying methods are introduced with dyeing and mordants to produce controlled elaborate designs. In this way the decoration is constructed rather than being applied by smearing or painting onto the surface of the fabric. Geometric shapes are created and color, shade and patterns can be built up from light to dark (Indianetzone 2013).

Batik

The use of wax, melted and painted onto a plain cloth, resists or stops any dye from penetrating and adhering to the fibers of the base fabric during the dyeing process. After dyeing the wax is removed by applying heat, for example with an iron, to the fabric that is laid between two layers of an absorbent material, such as paper. The melted wax sticks to the porous paper thus removing it from the fabric. This process is repeated as many times as is necessary until all of the wax has been removed. The resultant textile contains dyed areas and non-dyed sections where the wax was applied. Using this method of resist dyeing, multiple layers can be overlaid to produce highly complex types of designs. Different communities have developed distinctive traditional textiles as a result of refining these tie-dye and batik resist dyeing techniques (Edwards 2007). Indonesian batik, heavily influenced by Chinese and Islamic cultures, for example, uses three distinct steps in image creation, with an initial basic drawing, followed by the inclusion of motifs, and finally the background is decorated with traditional ornamentation (Situngkir 2008). Batik from the east coast of Malaysia, on the other hand, has developed in a unique manner due to its inaccessibility to colonial trade (Yunus 2012).

SYNTHETIC DYES

The Industrial Revolution increased both demand and production of textiles and, as a consequence of this, there was also a rapid increase in the demand for dyes. In 1856 a young assistant at London's Royal College of Chemistry, William Henry Perkin, was working on a project to synthesize quinine when he accidently discovered a process that produced a dyestuff that he subsequently found colored silk a vibrant shade of mauve (North 1969). After filing a patent for his invention, he went on to pioneer many new synthetic dyes and mordants. In an attempt to cope with the increased demand for aniline dyes, many scientists pushed forward their research into alternative colorants, producing an array of new synthetic colors (Pike et al. 2012). Over time and through prolonged engagement with their use, designers had found that natural dyes possessed unique features and characteristics that were subtle and gentle, while synthetic dyes were cheaper to use, easier and more uniform with a far greater variety of shades. Textile designers such as William Morris, who had started to use synthetic dyes, became increasingly disenchanted with the results of these new dyes; many believed that the results of their textiles were better with natural dyes because these produced harmonious colors that were more sympathetic to their images and patterns (Parry 2005). Synthetic dyes offered consistent tonal qualities, were less costly, readily available, colorfast, and as a result were increasingly popular. Unfortunately, dyeing with synthetic dyes also created a great deal of toxic liquid waste, such as bleaching agents and heavy metals (Mehta 2013). Mordants for natural dyes also created toxic waste, some of which still can be found in the sea floor near the places where rivers emptied out the effluents from dye/print works.

PIGMENT AND DYE CHEMISTRY

Pigment ink today is formed from inert, non-organic particles that are finely ground to ease their way through the print heads of the printer, although still solids and therefore capable of clogging. Dyes on the other hand, are made from organic compounds but tend to fade more easily. Originally pigment inks had the benefit of superior permanence, while dye-based inks provided a wider gamut. However, according to Johnson (2007) the biggest problem was the question of metameric failure, because even if two objects look as if they are the same color, under different light sources or conditions this is not always the case, and this affects pigment ink to a greater extent than it does dye-based inks. This is due to the dye's propensity to absorb into the fibers of a substrate, unlike pigments, which tend to dry on the surface because of the solid nature of the pigment particles. This can be extenuated if the grounds are not of consistent thickness, as this can result in a layering effect with uneven levels of light being reflected. As dye is water-soluble this is not a problem here, although some dye-based inks last a fraction of the time of pigment ones.

Research comparing inkjet printing and screen-printing on cotton found that each produced different results for color saturation and color gamut as well as tone. They found the inkjet-printed cotton had a better color gamut and quality of print but that to achieve color and tone similar to the screen-printed cotton they had to print three times. The benefits and disadvantages of inkjet versus screen-printing will be considered in greater detail in Chapter 7.

TOOLS

Tools are fundamental to our survival as a species, but also, without tools there would be no digital textile printing: no silk screens, no copperplates, no woodblocks, and no cave art. A practitioner's relationship with tools enables them to manipulate materials in an idiosyncratic manner (Bunnell 2004), and to visually express emotions, concepts and ideas. At first tools would have been the application or incorporation of materials, for example, sticks and stones, which were readily to hand and would extend the physical potential of the person using them. Tools can also be viewed as the means of imposing an idea on another material, such as a stencil. This action involves using a cutting tool that interprets, but also distorts, the idea of the image being cut out, as it is not possible to produce an exact match of that idea (Flusser 1999).

Cutting a stencil is a cutting, shaping, carving action that results in a physically cut out negative shape of an image that is to be printed. Both the material and the tools used have their own distinct languages, and when combined, this results in a new and unique language (Dewey 1934). The final print encompasses embedded, tacit knowledge from both the material and the tools that are used. The idea at the start is transformed, first into a stencil that is devoid of color. It is created using cutting, gouging or etching tools, in a material that is portable, malleable, capable of being cut but rigid enough to hold its shape. It must also be suitable for retaining the ink or dye to be used in the final artefact. Therefore, each material must be familiar to the practitioner so that their a priori knowledge can be used with skill to inform each step in the process, otherwise the initial idea will not be realized with sufficient mastery to make its production a satisfactory outcome. The aims and emotional responses of the practitioner are repeated through trial and error until the practitioner has become increasingly masterful in the creation and production of a personal interpretation of their ideas through the materials they are using.

The cave-based spraying action conducted around an actual hand produced a negative image from the hands of the practitioners, who blew or splattered the pigment solution at their hand against the surface of the wall. This process involved a tool that was readily available to them, allowing a negative shape to be created on the wall. In this way they were able to create an image by defining its negative, a process that lays the foundation for the development of textile printing in all areas of the world, regardless of cultural influences.

The stencil technique focuses on the negative of a shape and allows the substrate to reveal the image. This enables the user to decide in advance what shape the image will take and it only exists in the empty space of the stencil. Similarly a hand placed on the cave wall resists the pigment that is sprayed and, once removed, reveals the unsprayed, un-colored shape where no pigment is applied. This is exactly how resist dyeing works. The areas to be left un-dyed are either first painted over with a substance such as wax, that will resist the dye, or the fabric is tied or pinched in a physical operation to stop the dye from taking. It is the operation that dictates the shape of an image (Carden 2007). This process could then be repeated, and be further developed, similar to the digital creation of multilayered images on a computer.

Over time different materials and tools were used. Fingers applied pigment to walls, then sticks and brushes. Different techniques were developed such as blowing pigment paste from cupped hands or through the hollow bones of bird. Hands were used to create stencils and they were covered in pigment and pressed against the wall as a print. Gradually the range of processes and techniques grew. Multiple colors were developed, black from soot, red from iron oxide and yellow from ochre. These were mixed with saliva, oil, fat or honey, to produce different thicknesses. Solid blocks of pigment allowed controlled drawing to be undertaken over the surface of the cave. These were the basic building blocks that eventually led to digital textile printing. With no form of writing, every piece of learned knowledge that these early artists acquired had to be passed on from generation to generation, from elder to younger and from master to pupil. Social systems were developed for gathering and storing pigment. In the depth of the caves artificial light in the form of candles or oil was required. Scaffolding was devised and constructed in order to reach the top of walls and ceilings. Images began to be sourced from the environment, formed in response to conflicts in the surrounding area. During a period of over 10,000 years these visuals were re-interpreted and repeatedly worked over, on the same walls, by people who had never met. The cave art they created, however, continued as a testament to cooperative endeavour.

Nowadays a graffiti artist, such as Banksy (2006), creates a stencilled image on a wall. Such artists are, in many respects, similar to digital textile print designers because they create a sprayed image on a surface and walk away. The digital textile print designer also creates a sprayed, pre-designed image on a substrate. This can be displayed in a number of ways, such as indoors, or outside, on a building or as a garment or accessory. When someone wears a printed image it travels with that person. However, the designers of wall art are not present with their artwork, regardless of whether they are anonymous or not. Graffiti is stationary and its physicality only exists within the space it adorns. The garment sprayed with a digital image, on the other hand, moves from location to location along with the person who is wearing it. This means that the context and location in which a digital textile print exists is unlike that of the cave or building art because one is mobile, the other static, and each therefore impacts and communicates in a different way.

Palaeolithic cave painting involved hand-held pigment that was sprayed by blowing from the mouth. Gradually this technique evolved to hand-held cans of pressurized paint that are sprayed using air pressure as graffiti; both involve previously created images and stencil forms, each is sprayed using a stencil, a coloring solution and a source of pressurized air. The spray painted hand is further developed and evolved into tattoos, as a controlled, predetermined application of color and image on the human skin. This is somewhere between the static wall of a cave and the mobile garment that is easily and regularly worn then removed. Skin can be perceived as an exterior canvas of the human body. Other people view it every time we interact socially. As a mobile surface that could be decorated, it lent itself to increasingly sophisticated, elaborate and refined techniques. Like the walls of the caves, decoration would be applied by cut-out shapes, for example, the felt ram, by dipping them in pigment paste or solution and transferring the motif to the skin of the arm as a stamp. Shapes could be pressed down onto the skin either from a raised design, such as carved woodblock, or by an inked-up cut out. Individual dots allowed more intricate and controlled designs to be applied to the skin and this technique begins to resemble the matrix system that led to movable type and indeed the pixel-based imaging formation of digital images and printers. As soon as people realized that to replicate and control an image it needed to be capable of being clearly defined, they broke the form into its smallest common denominator, like in mosaics. Although each dot is separate, when seen joined together this creates the impression of a whole image. This is exactly how a digital textile print image is formed because each dot or pixel is printed individually to a selected, pre-chosen color, and applied to the surface of the base cloth. This is also how non-contact transfer printing works, and inkjet printing makes the whole process non-contact. Instead of a person determining how a print will be produced, including characteristics from their personal experience and physical blowing power, digital printing decides the pressure and coverage that will be produced. The printer is programed in advance so there is no human variation or personal skill involved.

MAKING

When artists or designers make something they begin with an idea of a final piece in mind, says Heidegger (1962); and he also believes that this determines how they are going to go about making the final artefact. Making requires skill, and Gardner (2006) and Sennett (2008) estimate the mastering of skill takes between five and ten years. As a skill is perfected, and the design process becomes more intuitive, the knowledge that is created becomes increasingly well embedded. Artefacts that are made can share many different kinds of meaning, but these only begin to be revealed when the viewer engages directly with them. Until this point, any knowledge they contain is tacit, not yet made explicit (Büchler 2006; Friedman 2008; Niedderer 2009c). It is worth noting that although tacit knowledge may be made explicit, underlying embedded knowledge cannot (Barfield and Quinn 2004). There is something particularly significant about being able to see evidence of the person who made an artwork in the fabric of the final piece (Dormer 1997); and to fully understand something it is first necessary to make a version of it. To fully understand the way it is made requires the practitioner to experiment, play and, possibly most importantly says Sennett (2008), make mistakes, because it is through taking chances, trying out something new and learning from their mistakes, that an artist or designer is most creative. As Eco (1989) explains, play and error with different materials and ideas result in new combinations that would not have been reached in any other way. Scrivener (2002b) and Schön (1983) also maintain that this state of confrontation changes the conversation they have with the materials and physical processes.

However, if digital systems are intended to provide new ways of making things, then another problem also needs to be resolved. As Polanyi (1974) explains, when we learn alongside someone who possesses a greater level of experience than ourselves, we unconsciously gather information regarding much that is not explicit, such as rules and knowledge, gained by emulating the practices of others. In this way it is also possible to acquire the knowledge of numerous hidden details that can only be gleaned by imitating another person. While a machine cannot work alongside and absorb knowledge from a skilled practitioner, another person can. This enables them to pass on at least some of the information they possess. People concerned with the act of making and using advanced technology will ultimately benefit if the hardware and software developers better understand how practitioners prefer to work, so that they in turn can include features in evolving new technology that skilled artists and designers would wish to use.

In attempting to address the unseen aspects of making, a dialogue needs to be created backwards from the digitally printed artefact to the numerous ideas from which it was created. According to Schön (1983), once a practitioner reflects on their practice, the work responds and the practitioner reacts to this: and, claims Dewey (1934), each discipline has a medium that sends out a specific message that is tailored to that medium. So while the practitioner reflects on their practice what is communicated is especially relevant to their particular methods of making. A challenge for digital textile printing is, however, that so many materials and techniques are integral to the process, oscillating between society and technology (Latour 1987).

Interpreting digital textile printing through Aristotle's (2008) four causes helps to explain the individual stages of production and the overall process of digital textile printing: the first cause, the material of its makeup, can be defined as the dye and the substrate or base cloth; the second cause, that which causes the idea of the image to be realized, is the coded image; the third cause, that which causes the coded image to be printed, is the printer; and, the fourth cause, the cause of the act of printing, is the printed artefact.

By reflecting on these four causes, it is clear that the first cause represents the materials of the artefact: in the case of a large format inkjet printer this is the reactive dye, including the alginate solution used to carry the dye to the print heads and the alginate solution required to coat the fabric prior to printing, plus the base cloth. The second cause describes what is needed to create the digital images: including the computer, the software, the connector cable and internet access. The third cause is the digital printer that enables the digital image to be transformed into droplets of dye and the digital dispensing of dye onto a base cloth. The fourth cause is the final digitally printed artefact. Therefore, I suggest the parameters of digital textile printing can also be defined using Aristotle's model as that which caused the coded image to be printed and the cause of the act of printing.

As is the case with an artefact, we can hold a digital textile print in our hands; as a digital file, we can send it anywhere in the world, including the full instructions for its production; as a digital image, we can view it as a light source on a computer screen; and, as an idea we can sense it in our minds. By attempting to imitate the idea, on a computer monitor, as a digital code or as a tangible object comprising fibers, dyes and the image as a digital print, we are also providing a version that can be turned over and seen from behind and we can learn to appreciate and experience it in many different ways. It is by making a physical version of an idea that we can learn even more about it. In his example, Plato (2000) explains that a couch comprises three versions: one is an idea of a couch; another is a carpenter's version of a couch; and, the third is a painter's interpretation of a couch that has been crafted by a carpenter. Plato also tells us that the painter can paint a couch without understanding how the carpenter has made it. This principle applies in a number of ways for digital textile printing, for a designer will not know the entire details of every previous process that has contributed to the technology or materials that are involved in its production. In much the same way as Aristotle (1996), Plato (2000) and Heidegger (1962) suggest, Peirce (1878) explains that an idea in one person's mind can never be exactly the same as an idea in another's. Not only is the idea unique to one person, but it cannot be replicated nor precisely realized in material form because a non-material entity can never be translated exactly into a physical material; when a practitioner attempts to translate an idea into a coded file it changes characteristics and again when the code is printed onto a base cloth it alters once more. The message from idea to code is different from the message from code to printed output (McLuhan and Powers 1989).

When Marx (1976) discusses the making of a coat, he looks at the labor that was expended from a human perspective, and explains that all such labor has a definite aim. He also points out that the materials used for making a coat, such as linen, have a greater value once they are associated with the coat than they would have as a piece of cloth. The coat has acquired embedded knowledge through the process of human labor, even though when the coat is old and almost useless as an outer garment it will still contain this knowledge or information. Similarly, for digital textile printing, the base cloth enters into an association with another commodity, the digital print; the fibers and dyes are, through the digital printing process, much more than the individual value of the components.

From a naturalist's perspective, Dewey (1910) considers materials as the mechanism for turning an impulse into an interaction with a person's surroundings. For example, using pigment on a cave wall fulfilled the basic instinct to make marks and must have been a satisfying experience otherwise the Palaeolithic mark-makers would have stopped doing it. Emotion, suggests Dewey, becomes intrigue and doing in action while reflecting develops a priori knowledge. So, if cave paintings converted experiences and impulses in the imagination into formed images on the walls, using materials and the anthropologic mechanisms (Marzke 1997) that Palaeolithic hominids already possessed, then there is an intrinsic connection between the message and the act of expression. The digital textile print should also be perceived as much from the materials and process of production as the final printed textile as an object.

Digital textile printing has also undergone significant experimentation and exaggeration, through its use of unrestricted color, repeats and imagery, and it has been incorporated into other areas of creativity and therefore has already gained a certain level of validation. Once practitioners are able to design and make with the materials involved in it, then a degree of anticipation and sensing of what will or can come next will enhance their practical knowledge (Peña 2010). However, as Polanyi (1974) points out, while creating it is a benefit for an artist or designer to never quite match exactly what they have in mind with what they eventually produce, because an element of surprise can help prevent outcomes from becoming flat or predictable.

So, perhaps the newness and novelty factor that is still associated with digital textile printing will enable it to retain an element of mystery for a while longer. As Frayling (2011) points out, in medieval times, clay was used to emulate tiles when slate was too expensive, and bricks were crafted to replace stone for the same reason; both processes started out as cost saving alternatives but with time grew to develop their own individualities and characteristics that were highly prized. Similarly, digital textile printing's role will probably change from a faster, cheaper and easier process, that emulates screen-printing, to a highly valued skilled discipline in its own right.

MATERIALS

Using the model of Deleuze and Guattari (1996) to explore the meaning of materials in relation to the creation of artefacts and designs, it would appear that the Palaeolithic cave artists used the materials of their environment to translate their conceptions, affections and perceptions (Leach 1976) from the world around them as well as emotional responses to their personal issues of conflict (Dewey 1934). Similarly, digital textile printing can enable a materials-led translation of the contemporary designer's response to sensations, perceptions and emotions that they can construct out of current ideas emanating from the digital landscape that we now find ourselves inhabiting and increasingly negotiating.

Just as early humans broke away from Neanderthals and began to use pigments and tools to visually express themselves in response to their environment, contemporary artists and designers are utilizing materials that reflect their own situation. Rather than reach for raw ochres and minerals, creative practitioners are using computers and online forms of communication to express what they have to say and to access the views that others have expressed about them. As social beings that exist and evolve in a physical world, they also need to survive in a digital landscape. They still require apparel and cultural significance, but what digital textile printing is able to provide is a mechanism for expressing emotional responses through a system that understands the language of advanced technology. In this way the pigment used on the cave walls is processed and refined to flow through contemporary inkjet printer print heads in order to translate our digitally coded images onto textiles that can be worn as mobile, removable surfaces on which we visually express our relationship to the digital environment.

In everyday life creativity finds an outlet or fulfilment through the handling of materials (Barrett 2007b), and the synthesis of the emotional and sensational responses to that physical interaction is evidenced in the relationship we have with the range or variety of objects around us. Various cultures embody materials with associations that differentiate between the progressive stages in their evolution. For example, during his travels Marco Polo (1997) noted that the trees from which leaves were taken and fed to silk worms, had a layer immediately under the bark, but before the central core, and this material was widely harvested for making paper money on the orders of the Great Khan. Planting white mulberry trees in order to grow money has an interesting ring to it, especially when the silk moth, the *Bombyx mori*, is only interested in its leaves, and nobody wants to eat the bitter fruit. The black mulberry tree, on the other hand, has leaves that the silk worm is not interested in eating, is not used for printing money, but has fruit that is highly valued for human consumption, especially in conserves. The materials, depending on their application or appropriateness for specific functions, embody different values. Alginate from seaweed is processed into a key ingredient for digital textile printing; it facilitates the flow of the dye to the print heads and is a necessary component of the paste that is applied to the base cloth prior to inkjet printing using reactive dyes; without alginate this particular type of image would not be printed or fixed. Cave people may have eaten seaweed if they lived near the coast, but there appears to be no surviving evidence to date suggesting they utilized it as a mixing or binding agent for printing. As Dewey (1934) points out, each material dictates its own way of being worked with, and this is largely due to its individual set of physical attributes and characteristics; but what Deleuze and Guattari (1996) add to the debate is that the making capacity for materials is context driven, and this too is a progressing system.

CONCLUSIONS ON TECHNOLOGIES, SUBSTRATES AND DYES

It was Clerk Maxwell who first explained that images are made of colored light, not outlines that are filled in. But, for artists and designers, creating a version of an image using pigment with tools on a surface, when none of them are made of light, is a challenging task. However, it could be argued that digital textile printing is possibly bringing us nearer to this goal than any previous techniques have allowed, and that the closest to date is the prehistoric negative hand stencil sprayed images on the walls of caves. The hand stencil images contain real evidence of the physical hand that was there at the time, its exact shape, size and position, in much the same way the foot impressions of young children are preserved in clay or plaster nowadays to remember and place the print at the time in the youngster's life when they were very small, at the same spot and engaged with the pressing of their own foot in the mold making material. Evidence of existence at the same time and space of the intention to make an image is a powerful emotional connection, which traverses through time and space; so the connection on an emotional level is not just with the materials and tools that were used to produce the image, but the non-material presence it embodies. Once a hand stencil had been produced, though, the sharp outline of the hand was also present, and recreating the shape of a hand by drawing with a stick of charcoal or a chunk of ochre would be a natural activity but would also lead down the path of making an outline shape and then filling it in with color, so starting to deviate from Clerk Maxwell's definition but possibly the only route available for visually communicating the shape of a hand, other than in negative form, with the tools and materials available in prehistoric times. If people had one color of pigment, usually either red or black, with which to draw and paint, then it is understandable that they would draw outlines in one and fill in the inside area with a covering of the other; this is effectively what woodblock printing achieved. With a palette of only two colors the options would have been limited to red, black, un-colored background and black over-colored with red. If we turn this around, it also means that the understanding of color as a whole concept had been divided into individual, distinct hues; in other words the appreciation and application of color had become digital even before red ochre was used to make colored marks on surfaces or in the Mousterian graves. Dividing color into distinct hues so that it can be reconstructed and built up to form artworks is how images were traditionally created; without the units the construction process could not take place.

For printing, the means of reproducing takes the building of an image with colored materials and tools a step further. The various actions and thoughts that go into creating an artwork cannot ordinarily be reproduced. The initial engagement with the materials and the trial and error, false starts, reworking, reflection and contemplation that all go towards making something cannot ever be repeated; once a correction has been made, something of that act is retained, either in the mind of the practitioner or in the audience or even in their relationship with the material in future. However, with stamping or founding, with a mechanical means of reproduction, the creation of an intermediary crafted medium can carry the original message to the final outcome. The order of the carved lines, or the depth and arrangement of scores in an engraving can be used to repeat the practitioner's actions imparted on the wood or metal plate on another material. Thus a stamp can contain the instructions and knowledge of what is to be transferred. When this could be done digitally, the storing of that information means it is possible to send it electronically to anywhere in the world, easily, quickly and cheaply, and it will not physically degrade over time as the wood or metal would have done. However, because a digital image can be of unlimited colors or any selection of those colors, as a light source, there is no longer the requirement to break color down into individual units in order to compose or construct an image. And, as there is no need to divide color in the manner that was previously necessary, the methods or reproducing the color is similarly not required to be divided in the same way either. So, for the first time, color can be used to create images of a photographic quality, by capturing and displaying images as colored light depictions rather than outlines filled with a limited number of blocks of distinct colors and this is why digital printing has changed the world of printed textiles.

4

THE PROCESS OF DIGITAL TEXTILE PRINTING

This chapter considers the importance of the physical properties of digital textile printing and seeks to identify how and where constraints can be used to maximum effect. The various ways in which the ideas for digital printing can be interpreted allows designers to take many different approaches to their practice. The digital printing of textiles embraces a wide number of processes, from traditional techniques to advanced technology (Carden 2011c). As a consequence, artists and designers are increasingly supplied with novel ways of joining and intertwining many different layers of meaning and narratives from the digital and the non-digital landscape around them. However, this vast range of possibilities also requires many diverse skills to be mastered, and the artist or designer needs to consider and balance the time it takes to become skilful in specific areas, with leaving other tasks to others to perform. Sennett (2008) explains it takes more than 10,000 hours to master a skill to the stage where it can be performed as second nature, so it is not surprising that practitioners usually find it hard to stop and think about what they are doing while they are

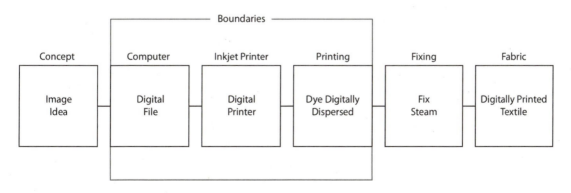

Figure I Stages of digital textile printing. Susan Carden 2010.

in the middle of creating artefacts. However, it is necessary for the tacit knowledge to be made explicit if the conducting of practice is to qualify as research, and under these circumstances it is also important for the artist or designer to document and record every aspect of the application of practical knowledge while they work in the studio setting.

This diagram shows that the designer is completely involved in the design process until the image goes to be printed by the inkjet printer. At this stage the involvement of the designer stops and the machine takes over. As Heidegger (1977) states, when the human hand relinquishes control to technology, we feel the need to control the machine. Eco (1989) even goes so far as to believe that we create machines only to be immediately oppressed by them. The reason is that the relationship we have with the non-human technology is so abhorrent that we incorporate aesthetically attractive features with the machine to pretend we are not enslaved by it. For, maintains Gardner (2006), human beings cannot be in charge of a situation until they have first conquered it; to work with advanced technology, a practitioner must first begin by mastering it (Carden 2012).

A key benefit for many practitioners and researchers is that a large format inkjet printer, such as the Stork Mimaki TX2, uses water. In digital printing this is not always the case, for example, dye sublimation printing involves solid dye that is heated until it becomes gaseous and is then transferred directly to the base cloth (Ujiie 2006). During dye sublimation printing there are few opportunities for interventions because if there is insufficient heat the process will not work, and if the temperature is too high this will cause the dye to spoil. Inkjet printing, on the other hand, due to the nature of reactive dyes, such as their reliance on water, provides situations in which it is possible to introduce complementary non-digital processes and techniques, and a number of these are explained in Chapter 5.

IMPORTANCE OF LIGHT

In digital printing, because of the importance of light, both for displaying the image and for identifying every single color that may be being used, especially if the image is photographic, means that creating traditional types of evenly balanced color-ways is not practical. As Joseph and Cie (2009) explain, the methods of production for digital textile printing determine what kind of design is created, and how and where they can be manufactured. For example, in digital textile printing the print heads are never in direct contact with the fabric, and the printing process cannot print white, so immediately we know that if there is any white required in the final design, then the base cloth must be white, and as the entire image is printed in one pass, there are no color separations and therefore no areas of multiple printing that would be expected from traditional screen-printing. The slight overlaps from each color that are evident in screen-printing and catch the light as slight raised edges, and help to further define each color from the other adjacent ones, is not present. Also, being able to design without having to consider the implications and constraints of repeats is a significant feature of digital textile printing. This leads to the practitioner enjoying greater levels of creative experimentation with color, composition and scale; and, because there are no registration issues, millions of colors can be graded and blended easily. The designer can then realize the image on the fabric as a strike-off and may alter or modify it quickly, returning and re-printing again, meaning that short-run productions are far more cost effective and ecologically sound than they were before, and this also significantly reduces the downtime of the machinery (Campbell

2008). However, there is always the possibility that the nozzles in the print heads can misfire, and if damaged, these parts are not simple to replace.

The technology of the print heads is responsible for the resolution of the printed image; a resolution of say 6000dpi has smaller dots than one of 300dpi, and thus produces an image with finer detailing and clarity. This also requires the ink to be produced at different viscosities to enable the print head mechanism to fire the correct amount of ink through at the right speed. The ink is further engineered to reflect its required colorfastness, stability and color gamut. The main factors that impact on the outcome of the ink being printed on the cloth are the speed of printing, the type and positioning of the print heads, the movement of the ink through the print heads and the way in which the ink is deposited and attached to the base cloth (Campbell 2008). The image is also influenced by the choice of fabric.

VARIABLES

The type of fiber used, its qualities of absorbency or wicking will all impact on the way the printed image interacts with the cloth. A single image that is printed using the same dyes at the same time on a consistent length of cloth, can still result in different visual outcomes. Although it is possible to print with millions of colors the system is not able to provide a like-for-like match with the colors displayed as light on a computer monitor, nor the gamut seen on the screen. The process of printing from a large format inkjet printer is nevertheless more accurate than the variations thrown up by natural fibers and the fixing methods, says Campbell (2008); so there are many aspects of the components and processes that alter the visual of the final outcome. He also notes that whereas textile designers are already prepared for these variables, artists and designers from other disciplines who can now easily create their own digital textile prints are not always so familiar with the idiosyncrasies of the system, and they do not have the prior skill-set or knowledge, so they may be able to create new designs but not ones that are deeply rooted in the culture of textile design. If the past is anything to go by, in time, digital textile printing will change to address these anomalies, but first there is another issue for color management to consider. The light produced by a computer screen is additive, whereas the color printed on a fabric is subtractive (Collis and Wilson 2012), in addition to which, fabrics do not absorb the same wavelengths and everyone perceives light in a different way. This means it is particularly difficult to color match the light source from a computer monitor with the color that is printed on different fabrics, using dissimilar dyes and non-identical printers. According to Collis and Wilson, the printer can print fewer colors than the scanner is able to capture and the printer can make colors that the screen cannot display, so even though in theory it is possible to print what is on the screen, the human factor and the joining of non-equivalent materials and systems means that this is not the case. When digitally printed fabrics are steamed to fix the colors, all acid, reactive and disperse dyes produce a wider gamut, so the color of the prints changes throughout the entire process. The designer chooses the colors they wish to have and the raster image processor software converts that selection into data for the printer and the colorant. When digital printing was first introduced, it misguidedly tried to emulate traditional screen-printing, but at a faster rate. Mass production, however, does not fully consider the opportunities provided by a system that at first sight appears to offer no restrictions or constraints.

INTERPRETING IDEAS

A digital textile print designer, who sets in motion the idea as an interpretation into code and then supplies the information necessary to produce the print via a large format inkjet printer, does so with the aim of making an aesthetically interesting outcome. The digital textile print, and its social and cultural context, is grounded in the past, and is output by machines and advanced technology that involves a non-material, virtual state during its creation, and realization. The context of the digital textile print is constantly evolving, as are its standards, aesthetic considerations and expectations, so the print and the printing process, and how they are experienced, are also constantly evolving. Calvino (1997) points out that contemporary culture is seen as being discrete, meaning that what was previously considered to be a continuous flow of ideas, such as creativity, is now increasingly being divided, measured and combined. A contemporary view of cave painting, for instance, could be that the pigment was divided into manageable chunks and grounds, the grounds divided into different consistencies of past mixed with saliva or clay, further divided into sprayable and smearable solutions for printing and painting. A number of these divisions would then be combined and, using tools like bird bones, sticks, brushes, finger-tips and hands, a fairly extensive array of coloring agents and applicators were produced. Today's digital image, digitally printed, has similarly brought the continuous flow of ideas back to a discrete process, although with a major impediment. The ability to combine a digital file, a non-material entity, with material that is not code, is a challenge. We can combine many materials, but joining different media is complicated because a translation process is required; the idea of an image can be converted into a digital file, and then it can be transferred to a series of impulses that regulate the colors of dye that are forced out as dots at a predetermined speed and in a preproduced viscosity onto a material base that is none of those. The concept of discreteness is part of the reason why cave painting was able to evolve into what was eventually digital textile printing, but digital textile printing has its own issues and challenges for contemporary designers in that they have descended from a culture that is, by nature, continuous.

Digital textile printing is the process of digitally producing an image with digital referring to either the method of printing or the format of the image, or both. Can someone digitally print a non-digital image? They must first make it digital. A handmade image cannot be digitally printed until it has been translated into a coded file; so, no, it is not possible to print a non-digital image digitally; but, it is possible to manually print a digital image. The base cloth must be non-digital, although the digital code can be sent in non-material form, such as a pdf, jpeg or tiff, but, even though it can be stored that way, it is only a textile once it is formed on a base cloth.

MEANING OF DIGITAL TEXTILE PRINTING

Textile comes from the Latin *textilis* (*Oxford English Dictionary* 1998) from the verb *texere* meaning to weave. By definition, digital textile printing encompasses a series of textile processes: fabric construction, image creation and application. Until the image is digitally printed onto a base cloth it remains a set of process-led instructions.

Print is a verb, a doing word, and is therefore time-encompassing; if people intend to print, this means they will print; if they are able to print, and have the knowledge, information plus the hardware and software necessary to accomplish this, then they can print; but, until the transfer of the image onto a base is being undertaken, it is not actually printing yet. Once it has been printed, then the act of doing the printing is complete for one or more reproductions of that digitally printed textile, and effectively it, or they, are digital textile prints regardless of whether one or many are printed. The knowledge and information provided for one is by definition, reproducible multiple times. However, although multiple prints can be produced, the ink, dye or pigment for each is not the same because ink, dye or pigment is a natural material – it may be defined as reactive dye, or pigment or dye-based ink, but this is a genus term and the actual particles and chemicals or atoms used for each print are different, that is, not the exact same atoms as used in the next, or any other digital textile print, because one atom can only ever be in one place at one time. The same is true of the base cloth that is used. Regardless of whether it is silk satin, fine cotton or viscose, the fibers in any two silk satins or any other fabric are never the same, because again, one particular fiber can only ever be in one place at one time. The fibers may be similar enough to be called linen, or silk or cotton or wool, but they are not capable of being the same, because they are not the same object.

MULTIPLE PROCESSES

The essence of a digital textile print is the result of a number of factors, including complex images that were not previously achievable, and designs that demonstrate a uniformity and exactness. Also, the joint processes, technologies, materials and non-materials used in the production of digital prints all inform the visual aesthetic of the final artefact. If the character of digital textile printing is durable, then it will be possible to convey it from one place in time and space to another. However, the transported essence must also possess the same group of processes, techniques, material and non-material resources that are required to produce the authentic digital print. Latour (1987) explains that until all the components of a system are perceived as one, the set of diverse resources that comprise it, such as the base cloth, coating and developing, cannot be regarded as a black box, because the various components can be taken apart, removed or adjusted and this allows the operator to open it up and make changes. Similar to the early Kodak camera, a large format inkjet printer still permits the operator to intervene in several ways, such as coating the substrate prior to printing, altering the height of the print heads and fixing the image-as-dye after printing.

While the camera evolved to become digital, with image input and artefact output possible within the one object, this is not yet the case for digital textile printing. A major difference between a digital camera and a large format inkjet printer is that, increasingly, the final artefact from the camera is communicated directly as a digital file, that is, as a non-material output. Whereas, the only possible outcome from an inkjet textile printer is a substrate that requires additional resources, both before and after printing, to form a final artefact. Comparing the potential for human inter-vention, the early Kodak camera and the large format inkjet printer share the greatest number of resources, in particular their coating, developing and fixing capabilities.

DIGITAL IMAGES

The image that is finally displayed on a digitally printed textile does not conform to Plato's (2000) description of a painting, nor a written text on a surface, both of which he says lack the animation of speech. Living speech is directed towards an audience (Rancière 2004), but a digitally printed textile can still be positioned and worn in a manner that enables the person wearing it to communicate with an audience of their own choosing. This is more than Plato's version of a painting can do, or at least provides a digital textile print with an alternative way of speaking. The image on the fabric joins the wearer or displayer, and together they project an impression that is not mute.

IMAGE AS SURFACE

Flusser (1999) describes an image as a significant surface that represents two intentions: the first is the perspective of the structure of what comprises the image, and the other is the view of the observer. Traditionally images were not unambiguous series of complex symbols, such as numbers, and historically images came long before text, but digital printing poses a dilemma here, because due to the technical production methods which are a result of applied digital texts, a digital image must contain a degree of denotative qualities that are connected to number systems at a structural level. There is a significant difference between a digitally printed image and a traditionally printed image. The idea or concept of the visual image goes into the computer so that it can come out as a digital print although, generally, the practitioner has little understanding of what goes on inside the computer. By inputting, the designer knows they are creating something that can be endlessly reproduced, and what is in a digital print is code that comprises history, culture, collective memory, all going endlessly around in circles. Foucault (1970) similarly describes an image as a sign that both represents and is represented; and he explains that one cannot exist without the other. For digital textile printing this holds a particular resonance, as first, the idea of the image is three stages removed from the final outcome (idea to code, code to digital image, digital image to printed output); second, the idea signifying can be viewed as the message which McLuhan and Fiore (1967) suggest is actually the medium, so the practitioner's idea is a message that is translated three separate times with each message representing a different form of communication; and, third, Kyttanen (2010) claims that using advanced technology means that the data form the object, positioned perpendicular to the message. This, I would suggest, leads to the conclusion that the data are not the message and so what is represented and what is representing lies in the dialogue between the viewer and the object, but not in the data that inform the object.

PERCEIVING IMAGES

Perception, says Arnheim (1969), is the mental impression of something; different disciplines and individual practitioners have specific interests and group traits. However, digital does not require as much abstraction due to its capacity to eliminate a number of previously understood restrictions, so can more easily go directly from found image or photograph straight to print without abstraction or modification, and this is a very different creative process than previous printing methods. Traditionally, color, shape, repeat and scale all had to be decided upon beforehand and these decisions contributed towards the final object. Barrett (2007a) suggests, however, that digital textile printing is the vehicle for an internal human idea of a digital textile print to be realized.

Within culture, the symbolic meaning of our experiences, says Beckow (1999), is understood not in physical artefacts, but through ideas and knowledge embedded within them. What gives an artefact such as a digital textile print worth is that some people know how to design and produce them; some people assign a value judgment to them; some make things with them; and, some buy items produced from them. A digital file does not demonstrate differences of, for example, shade or color variations across threads of a digitally printed textile, just a version of colors without real qualities built into the picture. People perceive objects, says Hyman (2006), by seeing the colors that make it up; so the outline drawing we often sketch is not real, it is a misinterpretation or re-interpretation of the image we are trying to realize, that is, unless we are intentionally adding the outline for another purpose or effect; so, we are culturally conditioned to create fake outlines. This is reassessed in screen-printing where solid shapes rather than lines are prevalent, and in digital printing the grading of colors is now made possible. As Clerk Maxwell noted (Mahon 2004), everything we see is color, it is not outline colored in; this is a convention that has developed independently.

IMAGE SEPARATIONS

Woodblock printing, copperplate printing and silkscreen printing have in common the individual printing of separate colors. A digital print, however, is composed of dots that can be selected from millions of different colors, and color separating a digital image into monochrome or a limited number of colors from such a vast array of choices can be achieved in several ways. One option is to separate the image into one of the four individual CMYK color channels (Rigley 2013), another is to interpret the image as a program and separate out the individual colors according to their RGB values.

CULTURE

What exactly is a digital textile print? As a physical object, when broken down into its basic constituent parts, it can be expressed as a collection of matter and forces, protons and neutrons, the same elements that were once star dust, and have been around ever since the birth of the universe, in one form or another, repeatedly recycled and reformed. In themselves, these atoms do not immediately or overtly express a digital textile print, so where do they fit into the being of a digitally printed textile? Atoms joined together by forces produce molecules of elements that can be combined to form materials that can be differentiated from one another, that is, can be identified and named as, for example, cotton or silk. The way these became a cotton plant or a silk cocoon takes us down a biological or scientific route, and within that there is a fascinating scenario of how different naturally occurring cotton varieties can be propagated to produce naturally occurring pastel shades, and why silk worms fed on a variety of different foodstuffs, such as avocado or privet leaves, rather than white mulberry, create alternative silk filaments, of different colors and textures.

However, the fibers that can be identified as wool, cotton, linen or silk are divided into categories, animal-sourced and plant-based. The wool grows on animals, and wild animal fibers would have been harder to acquire in sufficient quantities to develop any fabric-construction techniques, so the herded animals provided fibers, wool from sheep and goats, that nomadic peoples developed into fabric such as felt. Plants such as linen and cotton, cultivated for fiber and yarn production, meant that the peoples must remain nearby in order to tend and grow the crops in one location, so these fibers were explored by people who lived in a climate and geographical location that allowed cotton, linen and so on, to grow; for example, cotton is not native to Europe, so printing on cotton only started when cotton was traded with India. These crops thrived primarily in countries that went on to become synonymous with highly sophisticated textile development and trade, such as Egypt, China and India. Although not as significant in terms of textile history, or influenced by international trade, other fiber yielding crops thrived in localized areas, such as nettles in Scotland, especially close to the traditional Black Houses where nitrogen was in plentiful supply and nettles were processed to produce a rather coarse form of fabric similar to linen, or in South America where the Aztecs used agave fibers to create fabrics for basic domestic items, as it was plentiful, inexpensive and hardwearing (Somervill 2000).

COMPOSITION OF COLOR

How is digital textile printing impacting culture? Research conducted by Barber (1990) confirms that the construction of textiles, once developed by twisting or felting fibers, was not hugely different across or even between continents, but colors, including dyes and pigments, plus fixation technical knowledge and application, were region-specific, and therefore impacted on the evolution of dyed, pattern or printed textiles in each of these areas.

How could digital textile printing fit in here? Since its introduction the coloring agents used in the digital printing process are no longer confined to one area, and this mobility of access also extends to the base cloth, the image and the process. Therefore, the final outcome or artefact can be effectively produced, translated and shared across cultures, making it a valuable vehicle in the development of cultural connections. Dutton (1992) maintains that pigment and dyes were widely known and used for personal clothing and interior domestic decoration, both for everyday use and rituals, so were an integral part of culture and its development.

Cloth production (Barnes 1999) is one of the first technologies and archaeological evidence shows it emerged in the Near East and Asia as a form of protection, followed quickly by social associations. This was around the same time as cave painters were exploring the use of pigments in various forms and combined consistencies, in South Africa, Namibia, Spain and France (Perez-Seoane 1999). While the garments that were constructed from textiles, particularly for funereal purposes, could be used to indicate the social rank and gender of the deceased, the fabric itself, in terms of its designer or producer, was important as a valuable, non-fragile, lightweight and tradable commodity.

Dewey (1934) maintains that anyone who creates artefacts or designs takes pleasure in the experience of them. However, as people are influenced and conditioned by artefacts these have a significant impact on the cultures in which they live. Digital textile prints are made possible by practitioners, although these textiles are also responsible for the experiences of the individuals who use, enjoy and appreciate them within the relationships that are established and connect them with their social surroundings. It is through art and design that early peoples celebrated and passed on customs, and the design of the artefact has two distinct meanings (ibid.). First, the way the visual elements converge in a digital textile print is reflected in its cultural context. Second it has a functionality that helps to explain how and why all the material components of that design come together. The structure of the base cloth, the fiber of the yarn used to make the cloth, the dye that is applied to the yarn or cloth before printing, the pre-coating solution that is required to make the dye-based inks bind to the cloth, the reactive or pigment inks of the image sprayed through the print heads of the inkjet printer, the large format inkjet printer and the software needed to run it, the digital image as a coded file on a computer screen before it is rastered, a process that converts a vector image into a pixelated version, to divide it up for digital printing, even the fact it needs to be plugged in to an electrical source with the required fuses, so that it will run, are all determined by the culture and society and order of things in which we negotiate our lives.

The continuity of this system, moving from generation to generation through these groups and societies is what keeps cultures alive and is evidenced by the designs and artefacts that society produces. However, digital textile printing is so new that its impact is not yet clear; also, it did not just emerge as if appearing out of nowhere. Although we invariably talk about it being the result of new or advanced technology, it is only new because it is not the same as the technology people were used to up until recently, so advanced and new are relative terms. What digital textile printing does provide is not exactly a direct line of progression from rotary screen-printing, because even though it is termed digital textile printing it is in actuality a spraying action onto another surface of a predetermined, rastered version of a light-sourced image, and so is arguably more akin to a spray-can graffiti artist who uses a pre-cut stencil and spray paint to create their artwork on a wall. This is because the color is sprayed and there is no contact between the tool used to apply the coloring agent and the final surface, the image composition, scale, format and shape have all been conducted in another material from the one to be applied on, and the color, design, surface and tools of application are all brought together at the place where the final image will be created.

MULTIPLE MATERIALS

However, early textile designers created domestic fabrics that had all the materials, fibers and dyes, methods of production and skills and knowledge associated with it, well-known, mastered and able to be completed by the one person. Digital textile printing, on the other hand, now requires a vast array of processes, machinery, technology, software dye chemistry, fiber cultivation, fabric construction facilities and so on, and it would not be possible or practical for any one person to be a master of all of these. In this way, the society in which digital textile printing takes place must work with and acknowledge systems that are formed by links and relationships with many different areas and specialisms within society. It is this understanding of the context of digital textile printing that helps us to appreciate its cultural relevance. Another cultural concern regarding the creation of textiles is that while less and less is known or controlled by a designer, due to the technology becoming increasingly complex and the systems more diverse, the further away from the master-of-all trades they become, and the outcome is that more is either taken for granted or somehow not joined up, not left out, just not made overtly explicit. For example, the synthetic dyes are used by practitioners who may well be able to gather berries or roots and dye wool but who would be at a loss to know how to create aniline dyes if they were not available to purchase online from someone they may never have met.

COMPILING

Pre-alphabet people thought linearly. When describing something, they would include all the information they possessed about that particular situation. Freeland (2001) explains that when a primitive hunter drew what was on an area of ice, they would detail not only what was on top of the ice but also what they anticipated would exist underneath it as well. For digital textile printing, designers such as Russell (2013) and Rigley (2013) similarly utilize background information often collected from a range of sourced images held in a variety of stores. These practitioners are then able to electronically compose, selecting inspiration from a window here, a tool there, an image from another window, flick through a folder here, a Google search there, an idea from somewhere else sparked by a visual recollection, a quick internet search, a found physical image, and so on. What practitioners are now able to do is to create a digital representation from the pick-and-mix that represents the cultural landscape they live in. Designers are able to apply the same basic principles as their pre-alphabet ancestors who created an image using what they imagined in the forefront of their minds, combined with what they found and selected that was lying around them, and include all of it in the final image.

COMMUNICATION

In cave painting practice, those who took part in the process, both audience and participants, were provided with different types of information. Leach (1976) explains that there is a significant difference between the way people pass on information backwards and forwards. The spoken word is very different from an abbreviated text message and the conventions and outputs from digital printing will communicate very different kinds of information to the person who created them than they do to the client who purchased them.

After all, what is the content of a digital textile print? McLuhan (1964) would suggest that speech, for example, becomes writing, so the content of writing is speech; writing becomes print, so the content of print is writing; print is the content of the telegraph, so the content of the telegraph is print. In this way the content of the telegraph goes to print to writing to speech to the original through that generated speech. What of digital textile printing? The final digital print goes from the initial idea to image to digital file to digital print on to a base cloth. But, what is the content of the digital textile print? It is necessary to say that it is an action that transfers a thought of an idea or concept that is non-visual and non-verbal into a visual manifestation of that idea or concept.

FUNCTIONALITY

Whether the digitally printed textile is used to fashion a gown for the monarch, or as a curtain for a shop, is a matter of indifference; however, the gown could end up in a museum, while the curtain might make its way, eventually, to a dog basket. That is, we are formed, or informed, more by the nature of the medium of advanced technology by which we communicate, than by the content of that communication (McLuhan and Fiore 1967). Thus, we are shaped more by the nature of advanced technology that can print files digitally than by the message communicated by the digitally printed textile outcomes.

We view any new technology in a way we could not do previously, as the picture we have of it is conditioned by where we are not in relation to what is now older technology, and as every new technology is different, regardless of whether it is built onto an existing technology, then the content of what it communicates is also different and so, by default, is the content of the message it sends out.

All digital textile printing communicates a message that flows from the computer screen to digital printing, the transferring of a light-sourced image separated or divided by rastering into pixels or dots, with these individual dots printed in one pass, but containing previous knowledge and information, combined messages accumulated since cave painting on walls, painting on human body, painting on a removable covering for a wall, painting on a change of substrate; the image subdivided and printable from a single stamp of one component, to multiple woodblocks for multiple colors, to copper plate, to screen-print, to digital print. McLuhan also warns us that while any new technology may bring advantages to society, it can also have less welcome disadvantages, and he calls these the side effects, noting that how and where these may impact is unpredictable (1964).

ADAPTATION AND INCORPORATION

Amicable and mutually beneficial trade and cultural exchanges during the seventeenth to nineteenth centuries led to new fabrics, dye techniques, images, pattern influences and construction techniques between the West and India, Persia, the Ottoman Empire, Egypt, and North Africa (Sardar 2000). Even though artists and designers think before they act during studio practice, it is also the case that in certain creative situations there is an element of prior knowledge involved that comes out of nowhere, as if it was the practitioner expecting it or not necessarily being aware from where the action came (Schön 1987). For traditional products or objects, the artefact is the result of the working and reworking of the piece, correcting flaws and keeping in aspects that seem right within the context of culture. Digital textile printing similarly needs time to evolve on a cultural level; however, this is happening so fast and has arrived so recently, that it is disrupting the natural embedding-in process of previous textile design production systems.

The numerous generations and social influences and systems that would normally have developed through the cyclical process of trial and error, then corrections that over time are not repeated, have not had a chance to fully happen yet for digital textile printing. The digital print designer must learn the skill of applicable knowledge, so as to understand what an action does and why, before they can become skilled in designing with it; however, digital textile prints are already part of our culture because we produce, use and acquire them and therefore think about and consider them (Robertson 2013). It is possible for the image to be the message here? The medium, clearly, has a significant impact on the final aesthetic of the image and thus, also, the message. The meaning is also dependent on the selection of base cloth as this contributes to how the image communicates. Digital printing can be used to create multiple copies of an image, or a one-off, so there is a choice. To print means to transfer a predetermined image from one medium to another. This process requires the preparation for the image to be printed being completed before the image is transferred, with the process of transferring providing the practitioner with the opportunity to impart their own personal characteristics, combined with those of the materials involved, onto the final outcome. Each print is conducted at a different time using different physical pressure and sets of transferring material, just as any two hand-knitted jumpers, knitted from the same pattern and using the same batch of yarn, will never look exactly the same. Everything made from natural, physical materials, such as fabric and dye, is unique. Even the wool from two sheep will be different, cotton picked at different times of the day will be different and any two silk cocoons are always different. So, similarly, if a print is reproduced, every version is different.

LINKING OLD AND NEW

A digital textile print, like any text and image, regardless of the substrate does not mirror the world; instead, it reflects its source says Pollock (2008). What then are the sources of a digital textile print? These would include source or causes, such as ideas and emotions, knowledge and skills, message or medium, previous knowledge, implicit knowledge, tacit knowledge but also technology. The technology itself incorporates codes and dye chemistry plus the fabric producer's ability to make decisions about the provision from the source providers (Latour 1987). However, people from the same culture are able to interpret artefacts and designs in similar ways (Banks et al. 1989) and the skills that they develop and master are the same skills that Schön (1987) tells us join them to the past.

Where are the cultural roots of digital textile printing? In the caves of Europe, Indonesia, America, East and South Africa, early people experimented with and expressed their interpretation of the landscape by making marks on the walls that then built up to form an image that reminded them of something that existed around them or in which they saw something that looked familiar from their mind's eye. From the evidence uncovered to date, how the natural inclination to decorate using the materials and tools around us developed would have been similar to the way a textile designer develops skills with mark-making and use of color today. It would initially look like something they would recognize; and by repeating experiments with the same materials the processes became more natural and mastered. Eventually it was possible to let the creative process turn this process around so that instead of marks reminding them of something familiar, they were able to think of something and make marks that resembled, to a certain degree or level, what they already had as a picture in their minds. In this way, the hit or miss making and recognition of meaning from marks became marks skilfully and deliberately applied, with the intention of resembling what they had contemplated, thought about and wanted to draw or capture. Once this control was achieved, ideas could be expressed as rough marks with no color, then with color via pigment in a solution, then pastels and then both pigment hard and soft as controllable coloring agents. A shape with no color was to become a shape with color and then texture and shading with droplets breaking up the density of the colored areas, and the marks that were made could be painted or filled to form compositions and patterns. This describes both the practical development and production processes of cave art and digital textile printing.

SYMBOLISM

The symbolic nature of the cave paintings may be unclear, but the materials, methods of creativity and tools are known to us, and due to carbon dating, we also know when they were painted. The skills and technologies used were developed over 24,000 years and in contrast the skills and technologies of digital textile printing are less than two decades old, although how they evolved will help to explain their current cultural context and future potential.

The knowledge we have from pre-literate people is contained in the artefacts left to us. The interpretation of the images on the cave walls may not always be obvious to contemporary eyes, nor embody sufficient information regarding how they were used and what the symbols meant to those people at that time. The bird bones would have clogged with the ground down pigment particles they mixed to form a solution, and the sticks and stones they used to carve lines or form images would have been rough and uneven, and the smearing of pigment paste and spraying of pigment solution uneven, but the spraying of pigment solution over a hand against the wall would have been life-sized, exact and, because it was to scale and in negative, would have been both a reverse image and an exact size and likeness, and must have been quite a contrast to the stick-created drawings around them. By being able to print something by not painting it, but by contrast painting its background so that the hand stands out, is in effect, claiming that the hand was there all the time, and it was the background that was painted instead. If a hand was always there, what else existed on the bare walls that Palaeolithic people might have imagined were there? Revealing the human hand form must have been a revelation, especially to people who could not speak to each other and could not write anything down in the form of speech or the spoken word.

For digital textile printing, the printed image is like spraying with bird bones, a united process or technique, but whereas cave painters could also use carving, smearing, painting, line drawing, spraying and resist spraying, plus printing with hands dipped in pigment solution and applying them to the surface of the wall, digital textile printing initially gave a spraying option alone. Practitioners and researchers are currently attempting to introduce handcrafting skills into the digital printing process to achieve variations in the look and aesthetic of the digital textile printing process.

CONSTRAINTS

The constraints of any medium, says Hughes (1999), dictate what can, or cannot, be achieved. With digital textile printing, it is first necessary for designers to identify the limitations and then use their experience and creativity to exploit them. This is perhaps more complex than it sounds, because many of the types of base fabrics that are now routinely used in digital textile printing were originally produced, often over many generations and in different countries, for other forms of textile design. Therefore the weights and structures of these fabrics evolved to suit particular types of textile practice and both the cloth and the design process developed simultaneously; however, the fabrics used for digital textile printing have not evolved in line with the tools of advanced technology. Not only is the base cloth created independently from the requirements of digital textile printing, but, the designer of the digital print is almost never the designer of the fabric upon which the final digital image is printed. As Hughes points out, the computer does not make everything we need to do easy; but what it does do is to make a few things extremely easy, and by so doing, it quite understandably can lead the designer in a direction that may not always be the best for them, or their original creative idea. Until designers control the advanced technology, the computer and the inkjet printer can be said to dominate them (Heidegger 1977; Eco 1989). The practitioners need to handle and control the medium in such a way as to accurately produce their intended message (McLuhan 1964; Hughes 1999); otherwise the constraints of the situation in which they are designing, rather than the designer themselves, will end up controlling the creativity of the final artefact.

This is further complicated because designing for digital textile printing eliminates the need for physical materials within the design stage of the process. The idea of an image is produced, then translated into code, converting it from an idea in the mind of the designer to a digital representation, and the digital image is output as a digital print on a base cloth; however, neither the physical version of the image before it goes into the computer, nor the physical print that is output is exactly the same. This means that the message that goes into the computer is not the same as the message that comes out of it. And, by using Adobe Photoshop or Illustrator, for example, the message that is input is further altered according to the constraints of these software packages.

Constraints are obviously not new. Following the end of the Second World War shortages led to the rethinking of design due to the restrictions caused by the lack of skills and materials, and these resulted in a review of hand-making that raised design standards (Arthur 2004). Textile printing has evolved and been shaped by a series of technical and material constraints, such as is seen in woodblock printing. The wood from each type of tree can also be subdivided into outer softwood and inner hardwood for use with specifics of details as and when these were required. The wood carver's skill and ability to interpret the artist's image characterized the image content while the printer's mastery of the pigment, its handling and application onto the various substrates used at that time, all combined to form the aesthetic and expectations of the process within its cultural context. Copper plate, on the other hand, had very different technical requirements and therefore needed many different skills. This meant that the silversmith, rather than the wood carver, was better placed to pioneer the development of copper plate for textile printing. In both situations, the image was conceived by a textile designer for output as a textile, but was worked, albeit masterfully, by someone who was not a textile designer *per se*. Screen-printing involves the application of individual colors, in separate actions, onto a base cloth through a direct process of smearing dye or ink straight onto the fabric, rather than onto wood or copper first. The screen-printing technique differs from either woodblock or copper plate, even though there is still a need to balance the relationship, between the person who creates the image, the person who transfers the image to a silk screen, and, for hand screen-printing, the person who finally prints the pigment or dye onto a base cloth. The constraints of any one of these key components, such as the wood carver, the silversmith, or the textile designer, significantly impacts on the final aesthetic of the printed textile and means that each output is the cumulated skills of three diverse people.

Also, each process or technique involved in the development of digital textile printing has evolved according to its own set of constraints. These include materials needed for each method of reproduction, for example, wood, copper and silk screen; scale, such as size of possible repeats and registration; and, color, as well as, type of dye, availability, fixing properties and mordanting requirements. Every one of the essential processes of digital textile printing, however, is faster, cheaper and easier than the methods from which it evolved. So, designing for textile printing without these traditional constraints is perhaps the greatest challenge facing contemporary practitioners.

While it is called *digital textile printing*, the textile is not printed and as a result the constraints that would normally create boundaries for the definition and identification of a digital print are not present. What we have is a means of spraying a version of an original image, consisting of unlimited colors, as a two-dimensional or three-dimensional image. It is a digital image sprayed or directed at a base material. The only aspects that represent our own selected constraints are the choice of fabric and the intention of the final outcome's future use or application.

CONSTANTS

The only constant feature in any digitally printed textile is the image that is repeatable regardless of what base cloth it is digitally printed upon. Also, within any digitally printed textile, the fibers at any two points on the base cloth vary, and as do any two base cloths, even if they are generally described as belonging to the same type of cloth or even from the same roll of fabric. Natural fibers are inconsistent as they are produced at different times, in different places from different chemicals. Take for example, a sample of silk. The silk cocoon is produced by a silk moth that has eaten white mulberry leaves, although they can also consume leaves from other plants, such as avocado. Each leaf of the mulberry tree is different in that it has been grown on different branches of the tree, have absorbed different sets of nutrients and photosynthesized at different times – each leaf is also slightly different in size and will be at a different stage in its development when consumed; so, even within a single silkworm's cocoon, the filament will display variations at different stages along it. Thus, when the silk is unravelled, then dyed, either before, during, or after, the silk is woven into cloth, and the silk cloth is printed upon, every minute part of that silk base will be different from every other area, every part of the image connected to every segment of the base cloth will be different from every other part, even if the image may appear to look the same. Thus, again, the only constant that can remain, is the digital file that represents the image that can be reproduced.

CHARACTERISTICS

Digital textile printing is characterized by a reproduction technique that produces an image on a base cloth from code and dots. Every printed image from a computer can be defined as digital, and each rendering on a base cloth requires the image to be redrawn, meaning that each image has a start and end and it is not a case of preparing silk screens, woodblocks or copper plate engravings which create a value and expression of intent to produce multiple copies of the original image. A digital file can be personalized or altered with ease, and requires no permanent, expensive and time-consuming actions to modify it. In this way the file is devoid of the numerous characteristics that are displayed by a woodcut, for which one person created the artwork, another craftsperson carved the wood for each color, and a third skilled person, a printer, printed the image, each with a unique perspective and set of mastered skills reflecting the materials, tradition and cultural context of the work being undertaken.

RASTERIZATION

By reducing an original image to a raster, the artwork is broken down into vertical and horizontal coordinates that inform the display unit of what color is to be displayed and where it is to be located. This action enables the image to be coded so that the inkjet printer, or laser printer, knows where each individual dot is to be printed in relation to all the other dots. The digital nature of this breakdown results in each dot being like a jigsaw puzzle. Individually they do not communicate the artwork as an image. It is only once all dots are in place that the image is fully revealed. The raster process determines where each dot will be situated; the physicality of the printed dot, and the information governing its location, size, color, and position in relation to all the other dots, are unique. The dot must be fully described before printing so that the large format inkjet printer is programed, is informed where to transport the necessary ink in an alginate solution through the print heads and is propelled out of the nozzles with a preplanned force, at a preplanned height from the base cloth and to a pre-mixed consistency and dye or pigment recipe. The drops are extremely small, around 50 microns, so finer than a human hair. Paper is specifically developed for each type of printer to accommodate the moisture content, drying or warping issues, absorbency, sheen, opacity, heat, different chemical compositions of the dyes or inks; textile bases on the other hand, while constructed from yarns, being of very different interwoven techniques unlike nonwovens such as felt or paper, are intended to be more flexible and so are intended to create a different physical outcome. This has its own challenges, for example, supplying sufficient ink or dye to fully realize the image, but not too much that it oversaturates the area and blurs the image, and too little ink or dye can mean that the base cloth folds or bends when the image is broken up or fragmented at the edges of the folded area. Although the dye penetrates the surface of the yarns of the fabric's construction, certain dyes, such as acid, reactive or direct, all adhere to the structure of the fiber in different ways. Not only do the type of fiber and the kind of dye play a significant role in the visual aesthetic of the final digital textile print, but the structure of the base cloth is key to the image and the manner in which it interacts with the base fabric. The image and the base cloth need to be sympathetic to each other.

This section has explained the importance of light and variables in the process of digital textile printing. By interpreting the different aspects of meaning that characterize advanced technology, it has also highlighted the significance of rasterization and identified a number of challenges that this brings to dye application on different types of fabric.

Plate 1 Multilayered digital print. Susan Carden. Multilayered digital print. 2013. Photograph by Susan Carden.

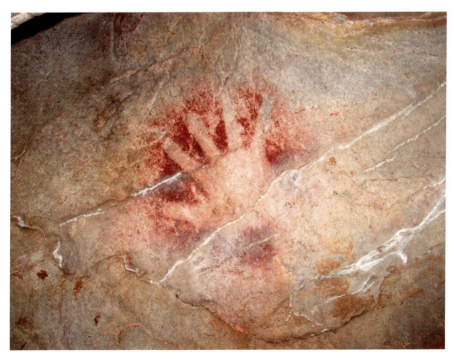

Plate 2 Detail hand stencil. Marcos García-Diez. Hand stencil El Castillo. 2015. University of the Basque Country. Photograph by Marcos García-Diez.

Plate 3 Bed curtain and valance of plate-printed cotton, printed and made by Nixon & Co, Great Britain, 1770–1780. Bed curtain and valance. 2015. © Victoria and Albert Museum, London.

Plate 4 Furnishing fabric with a design of fantastic flowers with pencilled blue. Furnishing fabric. 2015. © Victoria and Albert Museum, London.

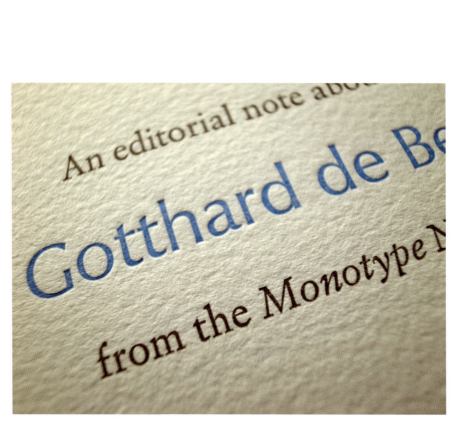

Plate 5 Letterpress. Christopher Wakeling. *Letterform 1*. 2013. Photograph by Susan Carden.

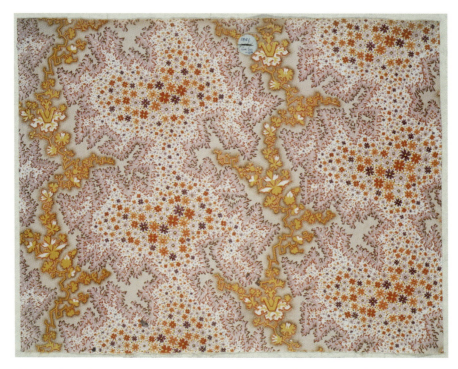

Plate 6 Furnishing fabric of roller-printed cotton, Lancashire, 1831. Furnishing fabric. 2015. © Victoria and Albert Museum, London.

Plate 7 Digital image on silk. Susan Carden. Silkscreen print. 2010. Photograph by Susan Carden.

Plate 8 Generated digital image. Alex Russell. *Cloth of Gold 1*. © Alex Russell 2013.

Plate 9 Laser digital print. Kate Goldsworthy. *Mono finishing-01*. 2009. Photograph by Kate Goldsworthy.

Plate 10 Make-ready sample in full color subsequently divided into the cyan layer and the magenta layer. Steve Rigley. 2007. Photograph by Steve Rigley.

Plate 11 Digital textile print. Steve Rigley. Sivakasi Textile Sample 1. 2007. *Journal of Craft Research* 4 (2). Photograph by Steve Rigley.

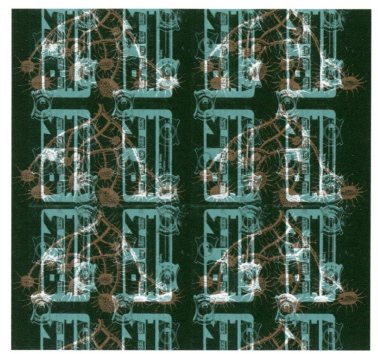

Plate 12 Digital textile print. Steve Rigley. Sivakasi Textile Sample 2. 2007. *Journal of Craft Research* 4 (2). Photograph by Steve Rigley.

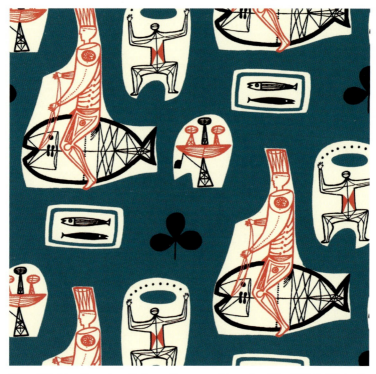

Plate 13 Textile print. Robert Stewart. *Macrahanish*. (1954) Digitally reproduced 2003.

Plate 14 Textile print. Robert Stewart. *Ardentinny*. (1954–55). Digitally reproduced 2003.

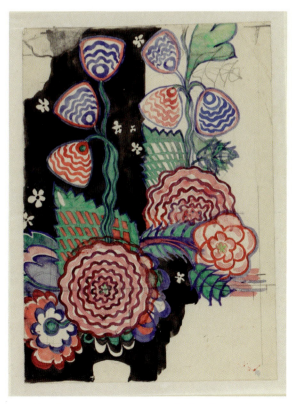

Plate 15 Textile design pencil and watercolor on paper 29.2 × 20.9. Charles Rennie Mackintosh. Stylized Chrysanthemums. © The Hunterian, University of Glasgow. 2015.

Plate 16 Digital textile print. Helena Britt. Assembly. 2009. Photograph by Helena Britt and Elaine Bremner.

Plate 17 Digital textile print. Elaine Bremner. Awaken Collection. 2009. Photograph by Helena Britt and Elaine Bremner.

Plate 18 Silkscreen-print. Susan Carden. Silk digital textile print. 2010. Photograph Susan Carden.

Plate 19 Illustration 22: Digital textile print on pleated silk. Susan Carden. Craft stitch technique. 2010. Photograph Susan Carden.

Plate 20 Digital textile print on silk. Susan Carden. Craft technique 2. 2010. Photograph Susan Carden.

Plate 21 Silk digital print. Susan Carden. Digital ikat on silk dupion. 2013. Photograph Susan Carden.

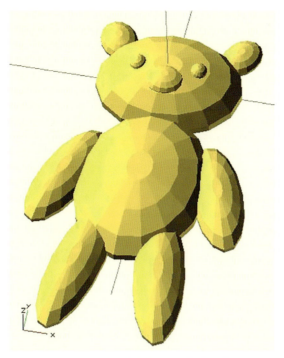

Plate 22 3D printed teddy bear. Scott E. Hudson. 2014. Carnegie Mellon University.

Plate 23 Print head detail, Scott E. Hudson. 2014. Carnegie Mellon University.

Plate 24 Silk textile print. Susan Carden. Silk dupion digital print. 2010. Photograph Susan Carden.

Plate 25 Silk digital print. Susan Carden. Pleated resist image. 2011. Photograph Susan Carden.

Plate 26 Silk satin digital print. Susan Carden. Transfer and batik image. 2011. Photograph Susan Carden.

Plate 27 Silk satin digital print. Susan Carden. Batik transfer image. 2011. Photograph Susan Carden.

Plate 28 Stencil image on silk. Susan Carden. Stencil image on silk. 2011. Photograph Susan Carden.

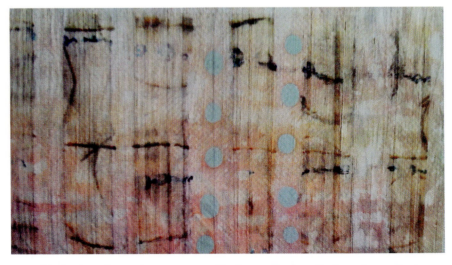

Plate 29 Batik and transfer image on silk. Susan Carden. Batik and transfer image on silk. 2011. Photograph Susan Carden.

5

ART AND DESIGN PRACTICE

This chapter examines a number of case studies involving practitioners, academics and developers who approach digital textile printing from a variety of perspectives. The range and depth of their work with advanced technology demonstrates the potential for digital textile printing to influence a large number of domains and to forge partnerships with neighboring disciplines. Reflecting on the interests and aims of the featured designers, this chapter also considers issues of sustainability, and acknowledges that there is a growing awareness on the part of designers, producers and clients, for the consequences of making when it involves technology, consumables, and materials, as well as the ethical considerations of the workforce. An example of a new, relatively sustainable technology is 3D printing, which is an additive manufacturing process and is explored in a number of the case studies in this section. This form of printing reduces the quantity of generated waste material because the product is formed by printing alternate layers of powdered material with a bonding agent, resulting in a system that leaves almost no residue. Another environmental advantage of this process is that the material used can be repeatedly recycled and reformed, while the data that constitute the designs can be virtually stored, remaining as files until required. This avoids the need for over production, reduces unnecessary transportation costs, and eliminates the requirement for additional physical storage facilities (Kyttanen 2010).

The practitioners and researchers include Alex Russell, a textile designer, educator and researcher with an engineering background; Cathy Treadaway, a textile design, educator and researcher with a ceramics and art creation background; Kate Goldsworthy, a textile designer, educator and researcher with a sustainability focus; Steve Rigley, a visual communications designer using digital textile printing to inform his exploration of narrative; CAT, the Centre for Advanced Textiles, a bureau printing service at the Glasgow School of Art for both internal and external clients, centrally placed to liaise with the Archives and Collections Centre at the Glasgow School of Art for reinterpreting and research-led projects involving the textile designs of former head of printed textiles department, and designer for Liberty, Robert Stewart; the sketches of designer and noted architect

Charles Rennie Mackintosh; and, exploring the potential of the archived material for extending textile and fashion design process understanding, Susan Carden, a textile designer, educator and researcher, using handcrafting processes within digital printing; and Scott Hudson, a human–computer interaction academic and researcher whose development of a new type of 3D printer allows woollen objects to be created with a digitally printed needle felting technique. Each of these cases use digital textile printing for very different purposes, and have been included here because of their links with current research models, highlighting the further potential for the outcomes of practice to reach a wider audience and generate further research investigations.

As artists and designers, digital textile print practitioners find creative and artistic expression through the handling of materials (Barrett 2007b), and this is a problem for digital textile printing because the handling of materials must happen before the image to be printed is uploaded to the computer and sent to the digital printer. The practitioner is unable to personally experience the materials, so the thoughts and feelings that this handling would instil in the artist or designer are not present. By creating images virtually and directly the output will not have been physically experienced by the person designing it. However, the computer provides artists and designers with a virtual space that is uniformly configured within relatively limited dimensions, routinely rectangular, uses light sources as colors, can never be handled directly by direct physical contact, cannot be viewed from multiple angles and means that much has to be left to the imagination or relies on previous expectations and a priori knowledge.

The practitioners detailed here have all been educated with, and have already mastered, the skills necessary for designing with digital textile printing at an advanced level. For younger designers and those new to the field, acquiring and becoming skilled in aspects that are less accessible may prove challenging. For example, it will still be necessary to understand how a traditionally printed textile is produced, with much trial and error, before deciding what an inkjet printer will be required to output in terms of dye and fabrics; how a certain base cloth will look and balance with particular images and will drape and distort and handle; the scales that will work best with images on those various fabrics when the designer is working on a screen that will undoubtedly be a different size from the intended final printed image or repeat pattern. What digital printing has done is to provide textile designers with a medium that will allow their images to be printed after being transferred into a format that enables them to be divided into a grid of pixels, and that arrangement of pixels can then be digitally printed as individual dots onto the surface of the base cloth. This allows us to print and reproduce the image in multiple forms on a variety of substrates in a variety of scales, as one-off strikes or multiple times, and store as the information to do so.

Beddard and Dodds (2009) claim that artists and designers have a unique way of working with advanced technology. The limitations of early computers, they claim, resulted in simple, distinctive geometric images that belied the complicated operations required to create them. Today, computers offer designers the potential to produce an aesthetic of greater complexity, while the tools themselves are increasingly simple to negotiate. However, just as a material has its own unique relationship with a practitioner (Dewey 1934), the computer also allows personal levels of engagement that McCullough (1996) compares to crafting. He maintains old processes are not forgotten; rather, they are being encouraged to evolve in a complementary way, so that a multilayered series of transformations is being created that helps to define what digital can achieve.

Benjamin (1936) explains that the mechanical methods used to reproduce artefacts alter an audience's appreciation of them, while Sontag (2008) suggests our culture is increasingly heading towards a state where excess takes precedent over sensory satisfaction. In order to fully understand an object in terms of both its method of production and sensory characteristics, Arnheim (1969) maintains it is necessary to engage with a multitude of perspectives on it. However, when one, or more, senses are excluded from this analysis, the picture is incomplete, and Sennett (2008) notes there is a balance that must be struck between acquiring hard-earned skill through repeatedly trying to master something, and becoming bored with the repetitive nature of going over the same thing again and again. On the one hand, if this operation is viewed as an opportunity for significant self-satisfaction it can be stimulating, but on the other hand, if it is not forward-looking, and anticipatory, it can be seen as tedious. This is a dilemma that routinely confronts practitioners making decisions about the pros and cons of digital and non-digital techniques, and finding which best suit personalities and ambition.

ALEX RUSSELL

Russell (2011) combines hand-drawn images with computer-generated transformations. His electrical engineering background is evidenced in the digitally produced floral patterns he creates, often informed by geometric principles. Yet, Russell (2009) has publicly stated that he misses the physical act of screen-printing. He voiced his concerns that contemporary design is increasingly about selection, and his PhD research explores programing skills as a methodology for investigating the digital design process. Although Russell's (2013) designs are primarily created for digital textile printing, he acknowledges the skill necessary for creating and making by hand, and routinely juxtaposes traditional techniques with the opportunities afforded to him by advanced technology.

Russell is particularly interested in the potential for digital textile printing to explore issues of repeat and color (Quinn 2009). By removing the constraints of both factors, his current designs are pushing the key components of digital textile printing, and are created by code, essentially creating designs to learn how to use processing to code better. Russell begins with an algorithm that outlines his idea or sketch and develops it from paper-based to working out his designs through the coding. His engineering training is particularly useful here, and his iterative process allows him to write and manipulate the images as code until he gets a design that works for him. Russell uses the code to create and manipulate, pushing to see how far he can go with a repeat while still holding on to a recognizable pattern. The structure of the design or composition is central to Russell's ethos here, rather than the element or unit within it. For him, the other fascinating feature of digital printing is that there is no limit on colors, so he enjoys the complexity, blends gradients and particularly likes photographic effects that the technology now offers to artists and designers. His *Cloth of Gold* (2013) design (see Plate 8) is produced as a generative design. The code models traditional printed textile design techniques, using cellular automata, producing an infinitely long pattern that never repeats.

CATHY TREADAWAY

Coming from a textile practitioner's background, Treadaway's (2006) doctoral research explains how digital textile printing can be used to support creative practice, providing artists and designers with a mechanism to more effectively communicate visual data. Her PhD demonstrates through live collaborative projects that this form of visual dialogue can foster further artistic practices. Treadaway also believes it is necessary for designers to understand how programers think (2004). The difficulty here is that it takes time, practice and patience to familiarize oneself with the skills and knowledge required and still further application to intuitively create with them. This is a challenging task in today's society, maintains Treadaway (2007). Her research also explores the importance of hand-making in digital and craft processes, and what it means to work in a virtual environment for those who are more used to making with their hands and physical materials. She looks specifically at the role of the hand in digital applications for creative practice during her collaborative project with Hodes (Treadaway 2012); this resulted in an exhibition of paper cuts by Hodes alongside a film documenting the making process. Using video and photography as research methods to generate data in a similar manner to Philpott (2011; 2012), Treadaway captured the act of making during another of her collaborative studies, Shoreline, with artist Bell (Treadaway 2012). Here she produced paintings, prints, photographs and textiles through the use of digitally created images. As a consequence of these joint projects, Treadaway has been exploring the implications of creating in a non-physical space amid non-global color fidelity. Her research identifies how practice in art and design is resulting in an alternative visual language informed by the integration of digital technologies (Treadaway 2007).

A virtual space poses a number of issues for practitioners who require emotional responses to be formed through their senses, because, as both McCullough (1996) and Treadaway (2011) explain, working with tools is necessary for all forms of creative making endeavours, but Treadaway finds that the reliance on digital tools, ones that lack fine pressure sensitivity, is in her opinion a major concern for contemporary practitioners (2009). Her project *Materiality, Memory and Imagination* (Treadaway 2009) investigates this relationship, between the use of the hands in creative practice and the physical material realization of their ideas. She studied three textile artists, Brandeis, Bell and Berbath, and during the study, noted how they all utilized the computer to select from their digitally captured stock of images, photographs, sketches and scanned pieces. One used the computer to produce multiple layers of digital images, another used it as a storage facility, and all three made use of it to review and select digital images. Through the shared imagery and sense of experience, color matching and processing were revealed as key causes of concern for the three practitioners and the distributed nature of the collaboration emphasized this fact. This was not surprising since artists and designers spend many years mastering their color skills and the variations that exist across systems have yet to adequately match the expectations and requirements of the practitioners using them. However, Treadaway has since created without color (see Figure 2), instead digitally printing a monochrome 3D rapid prototyping piece, *Sennen* (2011), thus working outside the realms of this color management issue. In this way the characteristics of digital textile printing, that include layering, multiple images, photographic quality, millions of colors and exact replication are deliberately circumvented so that she can focus instead on the form and potential of 3D printing to interpret her sketches from the Shorelines project. Interestingly, Treadaway's background includes ceramic design, so this piece allows her skill in a neighboring discipline to cross over into digital printing, just as Russell similarly uses digital printing as a vehicle for his engineering past.

Marshall's (1999) PhD also approaches advanced technology from the perspectives of craft and a designer-maker and, although focusing on architectural ceramics, he too demonstrates that digital technology can be used as a vehicle to extend the practice of artists and designers. It is not

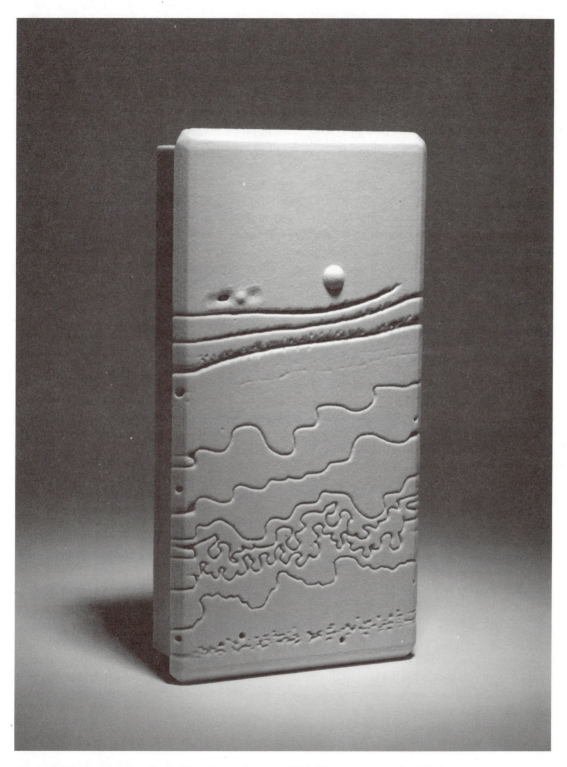

Figure 2 3D digital print. Cathy Treadaway. *Sennen*. 2011. Photograph by Dirk Dahmer.

only the digital printing process that increasingly provides novel opportunities for textile designers, photography, scanning, capturing stills from video recordings or the internet, intricate imagery and multilayering of digital images are among the creative opportunities growing in popularity now that advanced technology has eliminated the need for color separations, screens and a limit being placed on the number of colors that can be used.

Briggs-Goode's (1997) PhD investigates the potential of photography for digital textile printing and questions whether such complex visuals are appropriate for textile prints. Earley (2009) embraces the challenge and explores novel ways of incorporating photography as a method for importing, compiling, layering and manipulating images for textile printing and merges digital photographic imagery from different sources. Townsend's (2004) doctoral research, meanwhile, develops a new process for image capture, resulting in a novel way of engineering prints for garments and textiles. These examples demonstrate how practitioners and researchers are increasingly experimenting with alternative methods for image creation. As technology eradicates many of the historical restrictions by which designers were previously obliged to work, such as limiting the scale of repeats, which is explored by Bunce (2005), print bureaus are now able to output at almost any size, governed only by the dimensions of the base cloth.

KATE GOLDSWORTHY

Goldsworthy's (2013a) practice-based PhD re-evaluates how textiles are produced and re-manufactured ensuring continuous recyclability. Her research investigates closed-loop processes for materials, from up-cycling to continuous recycling, whereby materials ultimately need never reach landfill. Goldsworthy's interest is in identifying sustainable methods for production and finishing using new techniques such as laser technology (see Figure 3). This enables customized construction and finishing of complex textile outcomes that would have been costly and difficult to create with traditional processes (Goldsworthy 2013b). Although her aim is to produce infinitely recyclable textiles through up-cycling strategies, her use of new laser technology on a flatbed system allows short production runs and uses fewer materials and results in minimal waste.

As a sustainably aware designer, Goldsworthy creates from two basic principles: all materials should either be biodegradable or be able to be industrially recycled into new designs or products. Both routes require specific approaches and the ideal scenario is to keep the organic materials, fiber, dyes, devoid of chemicals of one side, and the synthetic materials, fibers, dyes and chemicals on the other (Goldsworthy 2013c). In order to prevent synthetics such as polyester waste from ending up as landfill, a process called closed-cycle polyester economy was introduced commercially in 2005 (Goldsworthy 2009); however, keeping the polyester materials free from finishing processes, chemical treatments and laminates, hindered its uptake. Concerns such as these are increasingly being taken into account by designers who design with the whole life of the materials and final product in mind; this helps them to facilitate the future life of each piece and determine where it may end up. If this has been considered effectively, the textiles will result either back in the ground as compost, or as a repeatedly recycled product.

Goldsworthy's designs incorporate laser-welding as a mechanical process for up-cycling non-woven polyester materials into resurfaced textiles (see Plate 9 and Figure 4); while the recyclable characteristic is maintained, the new aesthetics she creates are heavily influenced by

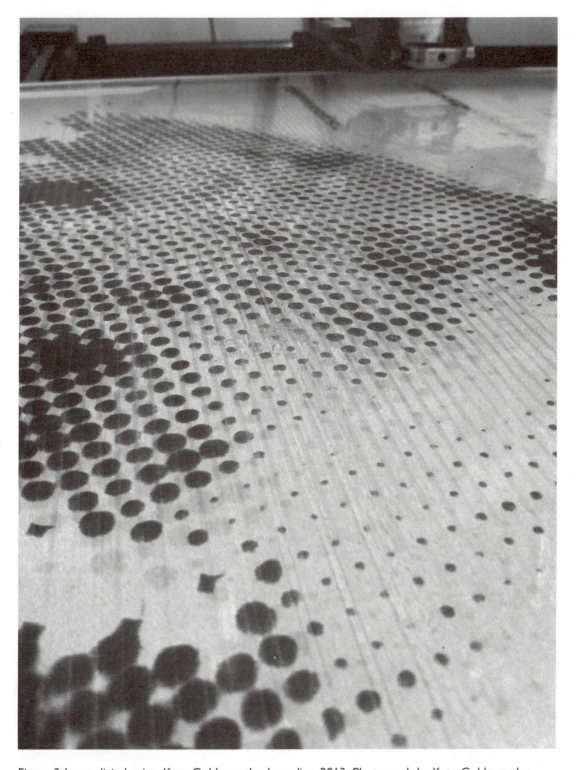

Figure 3 Laser digital print. Kate Goldsworthy. *Laser line*. 2013. Photograph by Kate Goldsworthy.

Figure 4 Laser digital print. Kate Goldsworthy. *Mono finishing-02*. 2009. Photograph by Kate Goldsworthy.

traditional hand-making techniques, such as lace-making. Instead of the hand working the intricate patterns formed by the skilled use of fine yarns and lace bobbins, a version of the complex lace designs is transferred via laser cutting onto the surface of the non-woven. It is not colored by the application of dye or ink during the transfer, but creates a stencil image, by laser cutting, with the color predetermined in the base waste polyester. The image is input as a digital file, and rather than inkjet printing the color, the process shapes the image in a cutting action. The cave painters cut their shapes on the wall without color, using the color of the base wall to represent the color of the final composition of the creative response to the environment. What Goldsworthy has managed to achieve is a similar approach of mundane fabrics, 50 per cent of globally produced textiles, and uses the tools that represent contemporary technology, to visually respond as a creative response to the material landscape around her. If the medium and the message are the same thing (McLuhan and Fiore 1967) then it would appear that Goldsworthy's message is the translation of her creative ideas imposed on everyday extremely ordinary fabrics through advanced three-dimensional laser cutting processes. The message here is: sustainability, recyclability, creativity and functionality.

Like Goldsworthy, Bowles is a researcher who addresses the negative impact of mass consumerism and fast design (Bowles and Isaac 2009). Bowles' research uses the digital and the non-digital to forge connections and online communities (2013). She helps to promote slow design and highlight the value of bespoke textile products by drawing inspiration from traditional craft techniques and advanced technology. Bowles collaborated with Newberg on *The People's Print* (Bowles 2013), a project that links creative individuals through online communities so they can share traditional skills and establish co-design opportunities. Artist Brandeis (2003) also explores the advantages and disadvantages of incorporating digital technology in creative practice; and, similar to Treadaway, she investigates the role that making by hand plays and the additional value it can bring to outputs of practice. Brandeis highlights the potential for complex imagery to be introduced and integrated in her designs alongside the increased speed of production and reduction in harmful waste chemicals. However, she emphasizes the lack of intimacy that results when hand control is removed from the creative process, claiming there is a lessening of color and surface quality in the final textiles. Also, Brandeis is fully aware of the cost of hardware and software, the challenges of color management (Campbell 2008; Treadaway 2007), and the time required for skill to be developed so designers are able to create effectively (Brandeis 2007).

Ujiie (2006), with a family background in traditional wax resist printing and a deep understanding of traditional techniques, embraces the benefits of advanced technology for textile printing, including short-run sampling and significantly reduced inventory. As with Brandeis, he similarly highlights the increased complexity of imagery that new digital printing processes offer, along with unlimited colors and the ability to eliminate any concern for registration. While Treadaway focuses on the role of hand-making in digital and craft processes for textile design and art creation, Brandeis weighs up the benefits and disadvantages of digital from the hand's perspective, and, Ujiie assesses the value of the technology for digital textile production. Campbell (2008), however, views these opportunities from an apparel designer's perspective, therefore looking for consistency in materials, maintaining that color management systems as well as raster image processor software result in the color on a monitor matching the final output. Campbell claims five factors affect the quality of any print: the speed, the movement of the print heads, the passage of the printing, how the ink is placed, and the height of the print heads (2008). The fabric used will also significantly influence the final textile, and he explains that many factors come together when creating digital textile prints. Campbell does, however, concede that the color gamut achievable with dyes, inks and base cloth cannot fully match that of a computer monitor, and when natural fabrics are compared to man-made alternatives, variations in pre-treatments and environmental factors, plus physical clogging of the print heads mean total consistency is not achievable. He does maintain that most textile designers are culturally prepared for these variations, and thus, he believes, they are willing to accept anomalies and work within the boundaries of these unpredictable irregularities.

STEVE RIGLEY

Rigley's (2013) digitally printed textiles are the result of rebuilding and creating new forms and associations from the creative destruction of observations and site visits to Sivakasi in South East India (Rigley 2004). He discovered waste sheets from commercially produced off-set-litho printing, that were used to check that the ink was printing correctly before the print run started, and were routinely printed over with unrelated print material from different, often diverse, brands, segments of typefaces and strangely placed photographs. The prints were formed from passes of tested individual CMYK runs, resulting in composite sheets of images. The random element that generates new compiled images is thus generative in nature with off-set-litho images cut and individual stages of the CMYK printing process acting as a quarter unit of the original design.

These single or often two color presses were of no use to the printer, generally A1 or A2 in size and printed on paper. So, the discarded as waste generated images provided a sustainable bank of images for Rigley to use for experimentation. He began by cropping and looking for visually interesting units of images, usually around 120 mm square. Digital technology, he says, enables designers to select from a wide range of imagery for cut-and-paste designs, particularly relevant in India where copyright is less rigorously enforced than in many other countries (Rigley 2013). So design studios openly encourage pick-and-mix and cut-and-paste from the internet, from recognized brands, from iconic photographs, and so forth. The traditional skilled commercial artworkers who were largely responsible for the press sheets that Rigley has collected over the last decade, have now been largely replaced by highly computer literate young graphic designers. The waste sheets evidenced the traditional processes and elements in cyan or magenta form (see Plates 10, 11 and 12). Digital technology enables Rigley to capture elements from the discarded sheets and rebuild images that themselves displayed a cut-and-paste aesthetic, thereby compiling and composing narratives. Rigley's unpicking required the image to be separated into cyan and magenta divisions, each representing a significant percentage of the original composite image but neither representing the exact shape or form of one color, rather the overall instance of cyan or magenta in the final image (clearly there are areas where pixels or dots will contain both cyan and magenta). This enabled him to take the composite apart in another form, into two divisions, which when joined again reconstructed the original image.

Advanced technology was then able to allow Rigley to overlay images and create complex patterns while at the same time emphasize their denseness, color balances and deconstructing qualities. He was conscious of the parallels with the Southern Indian town of Sivakasi, its deconstruction and rebuilding, and the cultural changes that have been taking place there. As he maintains, India has borrowed much Western iconography, supported and encouraged by the flexibility and opportunities afforded by digital technology and production methods (Rigley 2004).

Rigley followed a similar system when he realized the potential of the small, cropped areas he had gathered as a visual stockpile for later use. He acknowledges that the textile printing trade between India and the West had a major impact on the sharing of visual influences, technical knowledge and cultural developments; and, by electing to print his images on textiles, Rigley both responds to the historical importance of Indian visual culture in the U.K., and also the contribution Western visual culture has made up to the present day in India. Although, clearly, the relationship the West has had with India since the seventeenth century has been one of upheaval and destruction, it has simultaneously initiated rebuilding and the creating of new forms of creative expression and visual associations from the debris. This is a random element that generates new composite prints, like Russell's programed digital generative designs. These are generative images and narratives that are then brought together through advanced technology and digitally printed on fabrics – one practitioner uses the digital technology to generate and print, while the other uses traditional off-set-litho waste as a generative process and then digitally prints as textiles.

Many institutions within the U.K. offer students and external clients access to large format inkjet printers, for example the Stork Mimaki TX2, Epson 11880 or the Mona Lisa. These include the Centre for Advanced Textiles at the Glasgow School of Art; Digital Print Bureau at the London College of Fashion, and the Design Centre Bureau at University College Falmouth. There are a growing number of independent bureaus providing a fast, competitive service that includes advice and guidance to clients. However, with control of the printing process being taken out of the hands of designers, a further aspect of craft within digital textile printing is being removed.

THE CENTRE FOR ADVANCED TEXTILES

The Centre for Advanced Textiles, CAT, like a number of digital print bureaus, offers a service to both students and external clients. This service enables artists and designers to have their images or designs digitally printed from screen-based versions onto a fabric of their choosing. In this way clients can customize their designs, print images that contain complex photographic visuals and print on demand without the need to first sample or color separate their designs. They can also print on lengths of cloth as short as 1 meter to an unlimited maximum.

ClassicTextiles

CAT also offers a service they call ClassicTextiles; established in 2003 to accurately recreate a number of design classics of the twentieth century, this unique facility allows clients to order digital textile prints from a selected range of iconic work by designers such as Robert Stewart, Lana Mackinnon, Sylvia Chalmers and Lucienne Day. Stewart was formerly the head of printed textiles at the Glasgow School of Art, and designed for companies including Liberty and Donald Brothers, Dundee (Arthur 2004).

Stewart studied line-block printing as a student at GSA in the 1940s, and his work was highly influenced by designers such as Mairet (Schoeser 2004), whose deep understanding of traditional and synthetic materials resonated with his preference for design processes that he could undertake completely by himself. His designs, for example *Rock* (Arthur 2004), were produced as screenprints, but often included a version that was overlaid with a block pattern. Stewart was most at home when the designed image had a direct relationship with the constructed base cloth, but when the designed image had to become integrated as part of the construction process of the fabric itself, such as for his *Don Quixote* (Schoeser 2004) jacquard design for Donald Brothers, he was less comfortable, and this was evidenced by the reworking by the technical staff in Dundee who had to redraft the design to make it suitable for the jacquard loom's technology to weave. This is an example of a design that is aesthetically pleasing but, because it is intended for a different medium, in this case as a constructed textile design rather than a printed one, it is less natural for the designer, even one as talented as Stewart, and highlights the different specialisms and skills that exist within textile design as a discipline. Stewart may not have been as familiar with weaving as he was with print, but his vision and appreciation of color, shape and composition ensured that the designs he created, once aligned to the techniques of the jacquard loom, were as memorable and iconic as his printed textile designs. Digital textile printing, however, is now enabling designers from various fields to work in this particular area of textile design; but, as was the case with movable type, when less talented scribes were able to make themselves legible and thus work in the industry, making it easier does not always ensure it can maintain the skill of the original artform. However, Stewart particularly enjoyed the physical making of the image to be printed, whether this was in paint, or ceramics, especially in relation to the materials of the printing process. By the early 1950s, when the designs *Macrahanish* and *Ardentinny* were created (see Plates 13 and 14) hand-block printing was widely practiced while hand produced screen-printing was still relatively new, so Stewart's prints between 1951 and 1955 demonstrated his proficiency and skill with both technologies, having mastered the practice of one process while it was on the verge of being superseded by the other, and these prints showcase his ability in both (Schoeser 2004). Unfortunately, however, few of Stewart's designs created during the 1950s were attributed to him and only a small number are on record as being designed by him. The reason for this is that due to the popularity of British textile designs in the U.S.A. at that time, and in an effort to lift British manufacturing as a whole out of the post-war slump, individual designers were deliberately unacknowledged in favor of promoting a more general U.K. brand; as a result, Stewart's iconic designs are poorly represented in archives and collections from that period. Using the archives at the Glasgow School of Art, the CAT ClassicTextiles range was launched to recognize the importance and value of these designs, and to reintroduce them to a wider audience. Working with the full support of Stewart's widow, Dr Sheila Stewart, and his Estate, the team at CAT scanned, color matched and made digital files of the designs in the range, ready for on-demand digital printing of his designs on a linen union base.

This means the designs can now be obtained and made visible to a contemporary audience; not only were these designs unobtainable, but previously they would only have been printed if large quantities were involved due to the cost of setting up the screen-printing equipment, but with the new venture, pieces as small as a meter can now be printed on demand without the usual set up time, costs and inventory.

Mackintosh Re-interpreted

Unlike Stewart's designs, which had previously been in mass production, the printed textile design sketches created by Charles Rennie Mackintosh and housed in the archives of the Hunterian at Glasgow University did not made it into production (see Plate 15). The drawings remained at the preparatory stage and by the time of his death, had still not been completed ready for printing in any format, screen-printing or woodblock. So the team at CAT undertook the task of completing a number of the designs to make them ready for printing. For example, finalizing the colors and color balance, establishing the repeats and joining up the various loose ends and visual components of the sketches that needed pulling together (Campbell et al. 2008). Also, the scale of the sketches had to be decided upon, and for this digital printing allowed the team to make personal preferences, choices about what size to select: the process enables any scale with the only restriction being the width of the base cloth. So, scale judgment, color selection, repeat characteristics, and consolidation were applied to the sketches and made to work by the digital textile printing process; but, rather than make a fake version of Mackintosh's originals, the team took the perspective of using the advanced technology to produce a vision of what they felt Mackintosh might have created had he had access to the technology today. So, in this way, the collection uses the digital textile printing process to re-interpret his sketches and vision in textile form in a manner that has a contemporary edge regarding color management, interpretation and production values. Although Mackintosh's designs for houses and buildings, flower studies and landscapes, watercolors and interior designs, as well as furniture designs, are internationally known and widely reproduced, his textiles seldom made it to the production stage, so few people ever managed to see the sketches make into final textile form. CAT has enabled his ideas on paper to become, through extensive use of the wide range of archived material stored at the art school, an interpretation of how they could have looked on fabric had Mackintosh had access to the technology that is now installed in the building in Renfrew Street, Glasgow, that he himself designed in 1896, and saw completed in 1909.

AWAKEN

In 2008 the staff from Glasgow School of Art's Fashion and Textiles Department, the Archives and Collections Centre, and CAT, collaborated in a design-led project to investigate the potential for the substantial collection of material held in the Glasgow School of Art's archives to be used as a source for re-interpretation by the department's range of practitioners who demonstrate a number of significantly diverse idiosyncratic approaches to creative practice (Stephen-Cran 2009). Cross-fertilization of cultural associations can profoundly influence art and design disciplines, says Freeland (2001); for example, the print-making techniques of Japan were introduced at the behest of the Canadian government to enable craft workers to adapt their traditional designs for embroidered textile products into an alternative interpretation for enhancing the marketability of their original designs through printed artefacts. Similarly, the *Awaken* (2008) project at Glasgow School of Art encouraged a range of new conceptual possibilities from the original material stored in the archives, thereby forging connections between practitioners and across different cultures on a historic scale. Plates 16 and 17 demonstrate how Britt, a printed textile designer, and Bremner, a woven textile designer, successfully collaborated to use the archival resources.

Harrod (2007) suggests that digital technology allows designers to conceptualize for longer, spending less time physically making, and that this enables them to produce work they could not previously have created. Bunnell (1998) claims that advanced technology reduces laborious aspects of making, is more economical, makes more complex ideas possible, and encourages new aesthetic qualities. Both Bunnell and Harrod (2007) highlight the advantages of being able to create artefacts that were, until recently, unachievable due to the time it would take, or the complexity of the task. However, they and Eckert et al. (2010) make a case for certain constraints being beneficial. Eckert et al. claim that constraints on design involve the problem, the process, how this is achieved, and the evolving solution that shapes the opinions of the practitioner. They also explain that textile design is an under-constrained domain because, due to its manufacturing processes and materials, there are many opportunities for cross-disciplinary interactions. The under-constrained aspects of digital textile printing include no limit on the number of colors; the removal of repeat restrictions; the elimination of screens; and, an infinite complexity of images. This results in a substantially different skill-set being required by digital textile print designers. In this area, handcraft is less widespread and can be seen as a slow-design luxury (Fuad-Luke 2002). As the global demand and production of digitally printed textiles increases, due in part to improvements in pigment ink production (Provost 2008), this reduces the uniqueness of digital printing. As Quinn (2009) explains, companies such as Timorous Beasties are increasingly adding value to digital printing in situations where craft can offer a facet of uniqueness that digital, as yet, cannot. Utilizing the advantages of both offers a synergy that enables practitioners to work more creatively.

SUSAN CARDEN

The handcrafted techniques Carden uses in her textile design practice include a series of novel processes developed during her doctoral study at the Glasgow School of Art. One process involves preparing and experimenting with a quantity of alginate and urea solution suitable for coating a base cloth prior to digital printing with reactive dyes from a large format inkjet printer. This solution is normally applied to the surface of the fabric before printing, and the final print is steamed afterwards in order to fix the dye to the surface, and the residue of the solution is washed away. However, by painting with this solution by hand onto only those areas that are intended to retain the printed image, she is able to create an effect similar to batik, a type of resist-print technique, but in reverse. Instead of applying the solution to block the reactive dye, the coated area attracts the coloring agent. She then uses this method in conjunction with further dyeing techniques, including additional digital printing and hand dyeing, and produces layers of this alternative batik process to create complex digital textile prints (see Plate 18).

Another process involves raising the print heads of the inkjet printer to allow a thick hand-pleated and hand-smocked base cloth to be placed beneath them. By digitally printing an initial pattern she then unfolds the stitched areas to reveal random, free-flowing vertical effects of unstructured contours and lines (see Plate 19). This creates images that could not have been printed in any other way due to the complex color grading on and around the uneven surfaces of the stitched cloth. The raised print heads enable the dye of the image to be dispersed and diffused in a random fashion, and spreads down the three-dimensional contouring of the stitched fabric. The physical nature of the cloth and the distorting effect of the raised print heads combine to challenge the rigidity and exactness of the digital printing process. By combining techniques Carden is able to use the process to build up images and complexities in a way that was unachievable before the advent of digital textile printing.

A further process she developed to expand the potential of inkjet printing involves transferring a digitally printed image from paper to silk. She uses a natural liquid with a pH value of less than three to lift the dye of a printed image that has first been output to paper. The mobilized dye of the image is then conveyed from the paper to one of a variety of weights of silk base cloths, including chiffon, satin and dupion. By laying the silk fabric flat on top of a sheet of paper that has already been digitally printed, she then saturates the surface with the liquid, and seals the edges, leaving it to rest for an hour or so. When peeled apart, most of the dye-based inks have transferred from the paper to the fabric, leaving the pigment ink behind on the paper. This process enables Carden to incorporate natural phenomena that occur while the paper is wet, such as air bubbles and warping patterns, into the final image on the silk. Depending on the quantity of liquid, the thickness of fabric, the duration of contact, plus, whether the surface of the silk has been further manipulated by smoothing out air bubbles or not, the transferred images take on a variety of visual outcomes. Also, this technique can be used widely with cast-off waste prints, including some waste packaging material. These unwanted printed papers would ordinarily have been destined for recycling but this would have meant the inks were lost. Carden's technique has also been used in conjunction with the other techniques described above, and enabled her to create multilayered images using a variety of techniques.

Other handcrafting processes include cutting out a stencil shape on thick paper or card, placing it over the fabric, with a sheet of digitally printed paper sandwiched in between, spreading the liquid with a pH less than three over the cut-out shape, and leaving for half an hour. The stencil shape is a stencil-effect digital print. The transfer process forces the dye to penetrate deeper into the fibers of the base cloth, so the final image is more integrated into the fabric than a digitally printed image normally is, so the outcome does not give the impression of sitting on the surface of the cloth. When this technique is applied to a light open fabric such as chiffon, the resulting image is virtually the same on both the face and underside. In this way the reactive dye used in the large format inkjet printer does not require fixing after printing, it only requires a gentle wash to remove excess dye and liquid.

For another technique she combines a traditional batik process using wax to resist the dye during the paper to silk transfer process. In this way an image is created from melted wax on top of the dry silk, and the printed image on paper is transferred with the final outcome ironed between two sheets of paper to remove the wax. She then repeats this process with the first technique, the negative resist using alginate and urea solution, and digitally prints an image from the large format inkjet printer, so that the final textile is a composite of both forms of resist techniques (see Plate 20). A double-process version: one resisting the image while at the same time, the other attaching itself to the fabric.

By digitally printing an image on a heavy silk, such as dupion, she is able to remove the weft threads by gently pulling the individual ends. Once these have been taken out, she attaches the remaining warp threads, the vertical ones, to a frame and reworks the horizontal weft with a plain white silk yarn. This produces an ikat effect because half of the original digitally printed image remains on the weft and when rewoven the image is slightly disturbed and diffused, similar to a traditional single ikat process (see Plate 21).

These techniques were developed during her practice-led research and were produced using a methodology of grounded theory combined with practice. This meant she was reflecting on what she was making in relation to how she was making it, through processes, actions, prehension and transfer from tacit knowledge to explicit knowledge, rather than focusing on the emerging final artefact. Also, by intertwining practice with grounded theory, she was regularly suggesting to herself issues that might be informing successive substantive theories as they emerged during the data generation … It was only by being personally immersed in this practice that she could fully respond to the phenomenon and record directly from a first-hand account what had taken place in the studio.

SCOTT HUDSON

Hudson, a professor in Carnegie Mellon University's Human-Computer Interaction Institute, has developed a system that allows an object to be drawn digitally and then produced using wool yarn as a 3D structure (Hudson 2014). This new technique is similar to the existing Fused Deposition Modeling process (FDM) in that it layers and builds up the form of the physical object from a material that is fed or squeezed through the nozzles of print heads. However, unlike current thermoplastic 3D modeling systems, the process Hudson has developed creates a soft felted object from wool yarn or a wool rich blend rather than a hard plastic product. The yarn is extruded through the print head of Hudson's device in a continuous manner much the same as a melted thermoplastic does in the FDM system. A barbed needle repeatedly punctures the extruded path of the yarn as it is layered in a horizontal slice. The spikes on the needle felt the fibers together by a series of jabbing actions as they penetrate into the yarn and down through a felted foam base. By repeatedly layering and penetrating the yarn as it emerges in a continuous stream from the print head, the process builds up consecutive layers in order to create a three-dimensional structure. The final outcome represents a physical interpretation of the initial digitally visualized object.

Unlike FDM, in which the cooling of the melted thermoplastic causes each layer to bond with the one below, this new technique uses the puncturing needle to make connections along the route of the yarn as it is laid down. Because the object is built up from loose, twisted fibers of wool that have a natural crimp quality, it produces an uneven, soft and handmade look. The aesthetic is more akin to hand knitting or crochet than the rigidity and precision we associate with 3D printing in plastic, metal or glass. The developers have noted that wool works best, and it can be successfully produced with wool blends of up to 50 per cent animal fibers. As with traditional felt, the kinking characteristic of animal hair is key to its tangling ability, and so when the barbed needle is used, wool such as sheep's helps it to link and catch onto the yarn of the base layer better. This process is therefore similar to traditional needle felt making practices.

Multiple thin layers are produced working from a base upwards, although, as with other 3D printers, overhanging sections are not possible because each layer must bond with the one below, so the shape needs to curve either straight upwards or inwards as it is built. After printing, the sculpted form is peeled off, or cut free, from the felt foam base used as a starter, resulting in a flat underside. However, if a more rounded object is desired, then a second piece can be produced and the two joined along the flat sides. The teddy bear prototype shows the curved surface of a digitally drawn bear printed with wool yarn to create an object that is recognizable as a soft toy. It has a hand knitted appearance with irregular surface features that we instinctively know will be soft to the touch and instils the desire to both hold and hug. With the potential to interpret digitally drawn characters as springy, tactile and huggable objects, it is understandable why Disney Research is supporting this project (Disney Research 2014).

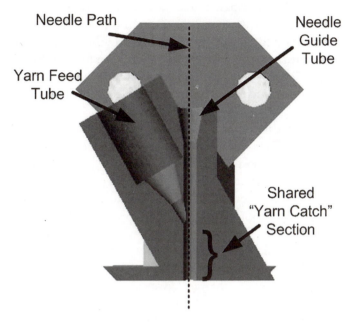

Figure 5 Feed head cut away. Scott E. Hudson 2014. Carnegie Mellon University.

At present the device does not cut the yarn that is used in a continuous length, so a post-printing stage is required so that unwanted lengths of wool, left between felted areas, can be cut away. Similarly, loops and twists that may occur from one needle penetration point to another can also be hand trimmed with scissors after printing is complete. This gives the designer an additional handcrafted opportunity in the making process.

Possibly the most significant advantage of this new technique is the potential to create rapid prototyping and customization of soft objects with a handmade aesthetic. For textile designers in particular, the fact that the process uses wool yarn, rather than a hard substance such as plastic, also enables them to work with a material that is both tactile and imprecise. This is a familiar way of working for textile practitioners, for whom the ability to personally respond to the natural give and take, bounce, elasticity, roughness or smoothness of different yarn fibers is part of the creative process.

Because the new felting printing process closely resembles the 3D or FDM system, as well as a traditional hand needle felting technique, this represents a fine example of how developers are beginning to take what advanced technology can achieve and make it work in ways that embody the hand skilled element that is fundamental to the way textile designers create. With this new technique practitioners are able to break away from the precision and uniformity of digital printing, allowing them to combine the benefits of advanced technology without losing the characteristics and workmanship of risk (Pye 1968) that making by hand can provide. This is an exciting opportunity for designers and allows them to extend the range of what solid modeling software and additive manufacturing with its prototyping, sampling and on-demand facilities, can offer them.

Its deliberate arbitrariness is a key feature in its appeal for textile practitioners. The technique enables a material to be applied to a base at the same time as it builds up a three-dimensional object through the printing process, so the printing action is both a textile applicator and textile constructor; that is, it prints and constructs simultaneously. In hand knitting the yarn is fed through the hands and constructed using needles into a textile, or for hand needle felting the roving is supplied through the fingers and felted by the needle; similarly in this technique the yarn is fed through the print head and constructed by the barbed needle, so its processes are extremely similar to those used by handcrafters.

Choosing to create a teddy bear (Hudson 2014) as an early object to prototype with this new 3D felting technique is appropriate for a number of reasons. Of all memories, short or long-term, those from childhood are amongst the most deeply rooted (Kawai et al. 2002) and accurately retained (Pasupathi 2001); this phenomenon applies to people of all ages (Habermas and Paha 2002). The narratives of our individual memories, including emotional, autobiographical and episodic, are particularly important because they uniquely combine to define who we are (Strohminger and Nichols 2014).

Non-linguistic artefacts (Daniels 2001; Wallace 2013), such as teddy bears, can be useful objects to help people of all ages recall their childhood memories (Winnicott 1971). These recollections represent long-term memories that are distinctive to each individual, they are charged with symbolism and have been shown to be the raw materials of person identity (Pasupathi 2001). Recent research in the U.K. and Germany, involving over 6,000 participants, shows that 50 per cent of adults still possess their childhood teddy bears, while 35 per cent admit to taking them to bed in order to help reduce stress and relax; also, 15 per cent of men and 10 per cent of women treat teddies as their best friends, sharing intimate secrets with them (Ahmed 2010; Holler and Gotz 2013), so this phenomenon is not restricted to any age group or specific gender.

Teddy bears and dolls are two of the most popular toys of childhood, according to Holler and Gotz (2013). They embody high levels of meaning, and because they are companions of childhood, they help us to reveal memories of adjacent activities such as play, being comforted, sharing secrets, as well as activities such as hand-making by, or for, someone else. So the teddy bears produced by Hudson's new felting technique also have the potential to stimulate a person's embedded desire to make something. In this way handling and experiencing the felted prototypes can prompt designers and those with previously mastered making skills to explore the possibilities of 3D modeling in a soft toy situation.

As Hudson has stated, this process encourages the introduction of integrated electric or mechanical components, circuits and devices to be included within the yarn or cavity of its production (Buechley et al. 2008; Perner-Wilson and Buechley 2010). Also, from a designer's perspective, wool yarn has the additional potential to be dyed in multiple ways in a variety of colors and this ability to decorate provides significant scope for future experimentation.

MARY KATRANTZOU

Since graduating from Central Saint Martins' Fashion MA, Greek-born designer Mary Katrantzou has built her reputation on digital textile prints. She plays on the computer in a painterly manner (Menkes 2012), creating silhouettes that are engineered directly through her digital textile prints.

Heavily influenced by her early studies and appreciation of architecture, her unique designs create silhouettes that are sculpted and constructed through her prints (Bumpus 2014). However, an issue for designers such as Katrantzou is that digital prints are easily copied leading to multiple fake versions being produced. Katrantzou's personal creative solution to this problem is to increasingly incorporate highly skilled handcrafted elements into her collections (Quek 2015). These include lace, embroidery and metalwork embellishments (Leaper 2014), such as her collaborative stump work for her Autumn–Winter collection with Paris-based Lesage in 2012 (Katrantzou 2012), followed by the 2014 Spring–Summer collection for which she used blown up images of Lesage's intricate hand embroideries within the digital prints themselves (Katrantzou 2014).

SUSTAINABILITY AND ETHICS

Against a background of increasing concerns about sustainable alternatives for change, practitioners, such as Goldsworthy (2013b), Rigley (2013) and Carden (2013a), are exploring novel methods for practitioners to use waste products in order to create new types of digitally printed textiles. One technique was developed that enables designers to make use of the dye and ink left over from discarded paper printouts from desktop printers and waste packaging materials (Carden 2013a). The process also reduces chemical waste by eliminating the pre-printing coating of fabrics that is normally required before digital textile printing takes place with a large format inkjet printer such as the Stork Mimaki TX2, and also the post-print fixing process by steam, thereby saving water, energy and chemical waste at the same time as reclaiming colorant that would otherwise be disposed of by incineration or the use of further toxic chemicals. Early indicators are that the process reduces capital cost per print, lowers energy consumption and creates an original, novel aesthetic that reflects the hand-making element of physically engaging with the making process, rather than passing all production elements over to the large format inkjet printer to perform. The technique is not designed to replace the function of the large format inkjet printer, but rather its primary aim is to provide an alternative method for creating digital textile prints that reuse spent dye and ink and create a craft-led product. This process avoids an overtly digital look, provides a system that is low cost, discrete in nature, and can be operated using pre-printed waste material and thus without the need for any hardware or software provision to be located nearby. As a craft-based alternative technique it offers artists and designers, small scale or large, with the ability to create digital textile prints that recycle unwanted and discarded printouts while reclaiming ink from a variety of existing printed substrates that would ordinarily end up as post-consumer waste. Initial samples were created using a mixture of text-based and image-heavy printouts, giving a range of visual outcomes. Later samples, involved selecting specific themes, colors and tonal qualities from the found materials thereby constructing images of greater depth and visual narratives. This extending the range of visual options available to the designer by providing the option to intertwine waste-paper images of random photographs, cropping and positioning them alongside new printouts that had been specifically created then output by the designer to add personal interest to the final compositions, A further benefit of the system is that there are four creative routes that can be taken: first, the printouts may be found, unwanted digitally printed matter, that can be used as they are; second, they can be digital images specially created and printed out via a desktop printer on inexpensive paper, then transferred to fabric at some later date; third, they may comprise a combination of discarded paper printouts plus newly designed desktop printer outputs that are collaged and then transferred to fabric; and, fourth, the images can also be stored and saved until later on at a time when the designer wishes to use them, then the image can be output on a low cost desktop printer and transferred to a base cloth. This latter system lends itself to the production of a range of ecological products in outlying locations, by different groups for a variety of purposes, and thus the process may also provide social benefits (McDonough and Braungart 2002). The technique, as it has been developed, is presently most suitable for use with different weights of silk fabrics. The liquid necessary to transfer the images is food-grade, cheaply and readily available, and with the option of combining the process with other craft-based techniques, such as batik, the possibilities are only just starting to be explored (Carden 2013a).

Contemporary designers are finding inspiration in diverse areas (Weil 1999). They are adding value through the application of alternative techniques and intervening by handcrafting in the digital printing process. As a result of this, they are designing with consumables, technologies and materials with ethical considerations in mind. By thinking about the life cycle of a product, whether it can be up-cycled as Goldsworthy has achieved with laser-cut polyester nonwovens, or created with the ecological perspective of the object in mind (see Rigley's discarded outputs of off-set-litho sheets or Carden's reclaimed ink from unwanted desktop printer and packaging waste), this can be defined by the terms of slow design (Fuad-Luke 2002). Material that would not routinely be recycled can now be considered as a valuable commodity for sustainable digital textile printing. As a low cost product, from both economic and environmental perspectives, all of the above new approaches provide the practitioner with ecologically and ethically considered design options.

CONCLUSIONS ON ART AND DESIGN PRACTICE

These case studies demonstrate that several practitioners have been exploring novel ways of creating images using digital textile printing. They have been experimenting with different surfaces, intervening in the printing process, incorporating handcrafting, and investigating the potential for found waste material to be reintroduced into the digital printing process. However, the non-standard, physical qualities of the fibers and fabrics involved in digital or non-digital printing mean complete consistency cannot be achieved. These issues, raised by Campbell (2008), show the dilemma for designers working with advanced technology, such as the goal of accurate, uniform, predictable printed outcomes, required for mass production, juxtaposed with the aim of bespoke, slow design for artists and designers who often have other agendas. While digital technology is able to foster several varied roles, including collaborative practical experimentation and process forming through connections, within textile design practice this is still largely uncharted territory.

6

ESSENCE OF DIGITAL PRINTING

This chapter defines the nature of digital printing as a means of identifying its essence. As society and technology evolve, so also will the essence of digital textile printing. Dewey (1934) maintains that the term *essence* can be open to various interpretations, but that artefacts are capable of conveying experiences through careful selection and refinement of what is fundamental in them. By taking a freeze-frame of the essential elements of digital textile printing we can identify which resources or components will most likely develop and therefore change over time. To do this, we can compare the inkjet textile printer with an early model of the Kodak camera. As the current digital camera is more advanced than the early version, it is possible to use the differences between these two models to predict which aspects of the inkjet printing system are most likely to evolve in the future. Gardner (2006) describes this as an interaction of three aspects: the practitioner, the discipline in which they are operating, and their specialist field with its associated rules. In this section key authors from philosophy and sociology are included so as to question established theories regarding the authentic nature of digital textile printing. By taking a theoretical position, opportunities are opened up that allow digitally printed artefacts to be examined in terms of their status and how they are experienced. This theoretical language provides an opportunity to progress contemporary debates in the field of textile design practice.

KNOWLEDGE

When Harrod (2007) suggests design students might be resisting advanced technology by experimenting instead with traditional materials and techniques, there may be another, underlying reason. While she maintains some students and design practitioners are reluctant to engage with new technology, others, including Chang (2013), Russell (2013), Krogh (2013) and Fox (2013), are creating new ways of integrating the tangible conditions of materials with advanced technology (Braddock Clarke and O'Mahony 2007; Quinn 2009). As Gardner (2006) and Sennett (2008) point out, the length of time it takes to master a skill is estimated to be somewhere between five and ten years, meaning it is impossible for most students to be experienced enough during their undergraduate, or even graduate, studies to use advanced technology at such a high level.

KNOWLEDGE IN DESIGN

It was Aristotle (1996) who said that in order to gain knowledge it is first necessary to understand why something is the way it is. He also stated that in general, human beings either imitate what nature provides us with, or take what is already out there and add to it in accord with the four causes. Digital textile printing does not claim to imitate nature although it does appear to be aligned to the latter (see Chapter 9).

Knowledge in design can be explained from a sociological perspective (Scrivener 2009), with the outputs of creative practice communicating knowledge in a variety of different forms. For example, a digital textile print is evidence of its means and method of production; it also visually communicates knowledge about how it was made, by whom, for which audience, and using what materials.

KNOWLEDGE IN DIGITAL TEXTILE PRINTING

People who live, work and interact within specific environments have a natural sense of what things mean, and how they are undertaken (Dreyfus 1972); however, a machine, such as an inkjet printer, does not possess these skills and is unable to anticipate what to expect in any given situation, so for any device that is programed to do the job of a human being, for example print an image onto a fabric, it is not possible for it to learn or acquire relational knowledge from this operation, because that would require an inkjet printer to have a body like ours and a presence in the world like us too (Dreyfus 1972).

This inability of an inkjet printer to reflect and further use knowledge acquired from past experiences (Dewey 1910) to inform further actions is a major area of concern for many textile designers, because the practitioner has to decide early on in the design process, before the base cloth is printed, the exact instructions that the inkjet printer will be required to follow. Those decisions are taken out of the practitioners' hands at a far earlier stage than they would normally be for people conducting the screen-printing themselves, in their own studio, to their own designs; in this case minor tweaks or changes in aesthetic decisions can be made, or even added, when the colorant is being applied to the surface. Physically engaging with the materials, a key source of acquiring procedural knowledge and experiential knowledge is not available during the act of digital textile printing. This form of personal interaction with the materials of creative practice was, however, available as a matter of course to cave painters, and was communicated through a system of knowledge transfer at a social level and was made available to others through direct observation, demonstration, display, practice in action, and master-pupil relationships. So that, even though there were no books, no textual record of what was being created, how, or why, the visual documentation was in the artefacts on the walls and provides us today with evidence of how these paintings and artworks were made, although not the other forms of knowledge that were communicated to others at that time.

Practitioners are, now that digital textile printing is almost two decades old, questioning how it can embody what they would ideally wish it to, rather than accept limitations. Being generally educated to question and seek new ideas, new knowledge and new associations, this is understandable and natural. The cave painters did not stop; they tried out new substrates, such as skin and then detachable fabrics; new dyes from plants, animals, and then synthetic chemical dyes; new tools, such as woodblocks, copper plate and then silk screen; so it is only natural that digitally printed textile designers do the same. The inkjet printer has changed how textiles are printed, created, used and communicated, in a short time. As the cave painting knowledge was not documented and recorded it had to be continuously communicated to keep it going. Cave people painted with brushes and fingers, carved with sticks and stones, and applied color, used blacks and reds, thick and thin sprayed through bones and resist over shapes, so even with charcoal, red ochre, stick and bones, by themselves, they were able to draw, scratch, paint, smear, blend, spray, stencil, do negative and positive images, use no color, with two colors or one color, blend the two colors; so, either zero or one, two or three colors, use shading, various mark-making techniques, contact and non-contact processes, and that was with the very limited materials and techniques available to them.

Similarly, a digital image, colors of dye, an inkjet printer, a selected base cloth and an audience, today has other constraints to face: the print heads spray at a uniform pressure and uniform viscosity, as a uniform covering, so will take the color at a uniform rate. The image is pre-created or image captured, can be as varied as the practitioner wishes; so, is of limitless complexity, using unlimited number of colors. The uniformity of application, non-contact and predetermined, removes the human touch from the physical printing process. The translation of the light-sourced image onto a base cloth never can, and never will, match the image on the screen, for reasons that are non-changeable. So what we really can achieve is a wider complex range of images than ever before possible: colors never possible before; scales of repeats, or no repeats, never before possible; transferred, stored and sent as digital files. But these digital files, digitally stored as code and digitally printed are not made directly from the materials of the surroundings and applied on the materials of the environment to interpret or express an emotional response to the environment because the digitally printed textiles are a crossing over of materials, from physical to non-physical, the realization of where our culture exists now, one foot in the physical world and one in the digital world. The landscape is a mixture of online presence and activity while sitting in our physical homes or offices wearing physical garments that can increasingly display physical prints of digital images, so what we create quite rightly reflects our surroundings and comments or communicates our emotional existence within this world.

KNOWLEDGE IN DIGITAL PRINTS

Nowadays, we have the means and wherewithal to record, document and access in multiple forms of media. Both visual and textual knowledge, as well as digital textile prints, can contain tacit knowledge and previously developed knowledge from the vast array of people whose skills and knowledge is contained within them. Similarly, while the cave people made and left visible artefacts, the knowledge communicated and handed down was restricted as to its dissemination and so the knowledge it generated is largely lost to us. Exceptions are what can be gleaned from the artefacts

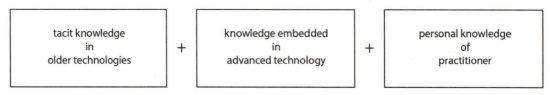

Figure 6 Knowledge in digital textile printing. Susan Carden 2013.

themselves and the workshops they left behind, for example at Blombos, as archaeological evidence. The creation of digital textile prints, on the other hand, relies on a significantly different range of knowledge bases and skillsets. This is because the dyes and inks are produced using multiple skills and knowledge through the refinements, mordants, fixing agents, pre-coating chemicals, pigments refined for print heads, color management considerations, produced in solution for flowing to print heads, mobilization, and viscosity, and so on.

The substrate, or base cloth, also embodies multiple types of knowledge and previous experience in the construction of cloth, the fibers and processing, gathering, cultivating, transporting, combing, bleaching, finishing and pre-coating. The image also is the combination of multiple skills, knowledge of image creation, composition, cultural context or considerations in relation to end use, for client, for market, for future embellishment, for apparel and so on, for image input, repeat or not, for scale and so on. The technology, the inkjet printer, is also the result of multiple technologies and the skill and knowledge from all the previous contributors to this advanced system. Also the documented, recorded, filmed, archived artefacts and technologies, natural dyes, synthetic dyes, inks, pigments and so on mean that the knowledge-base for the contemporary digital textile print designer is vast and yet actually creating or printing is or can be done without the designer even being there when the digitally printed textile is printed.

KNOWLEDGE TRANSFER IN DIGITAL TEXTILE PRINTING

So, the transferrable knowledge is not communicated to another individual. The main reason appears to be that the inkjet printer, as Dreyfus (1972) and Turing (1950) pointed out, is not human and cannot behave in the same way as a person does, that is, it does not, and at present cannot, think for itself; what it knows, or can do, is what it has been programed to do. It is restricted by what we have made it able to do, so it potentially can do all the previous things or knowledge which we have been able to do, but it is not able to reflect upon this and come up with anything new or novel that it was not constructed to do. Like a discovery or invention, a machine is not capable of recognizing eureka moments. So, the dilemma for us today is that we can produce a machine that can print limitless images with more colors than the human eye can see, or distinguish between, but the technology that produces it cannot teach anyone else and transfer the tacit knowledge or previously contained skill or knowledge so that a person can learn from it. We can watch it; we can read all the manuals and all the research published on what it does, and the outcomes it produces – but as for learning by watching, not yet. The machine, let us remember, has had all its previous knowledge input into what it is now. Knowledge is cumulative: each piece of new knowledge builds on top of all previous knowledge. So to look at, for example, an inkjet printer, what we see is a collection of metal, plastic, material parts joined together and plugged into a power supply; not only are all the forms of previously acquired knowledge different formats, involving alternative media, sensations, emotional responses, inventions, and so forth, they are by very varied people, the materials vary, the descriptions vary, the knowledge is a combination of translations from verbal communications, pictorial communication, non-verbal, sense-based. As Einstein (1987) said, mathematics is not easily communicated in text format; so also this knowledge is fairly vast in quantity, so practitioners generally use it without fully understanding or appreciating all the previous knowledge, tacit or procedural, that it contains.

TECHNICAL KNOWLEDGE

To acquire technical knowledge means to produce directly using the materials, skills and tools required (Nimkulrat 2007). A cave painter, for example, could achieve this by working and learning alongside another cave painter. However, largely a digital textile printing designer has different knowledge types to engage with, and learning by working alongside all the people and equipment and technologies that feed into how a digital print comes about is not simple, nor practical, and often impossible, especially as the inkjet printer usually prints out the artefact when the designer is not present. The computer file stores the instructions as code, which the designer cannot see or access; the image is a coded version of the light-sourced version of the visual version of the initial idea that the designer had, that is, a third-generation version of the designer's original idea – but to accurately and comprehensively convey the knowledge contained within the digitally produced artefact also requires technical knowledge for making with code, files, juxtaposing base cloths and dyes with light-sourced images. The characteristics of the individual brand of inkjet printer with its appropriate dyes and how they interact with the fabric of the base cloth and the pre-coating and fixing processes and color management of the systems used, plus working directly with these considerations alongside a skilled person who is an expert in all of these, particularly as most are systems and therefore non-human, is problematic.

While cave painters appear to have had considerable time to evolve and develop their skilled use of materials from their vicinity, the digital textile print designer of today needs to change and evolve far more quickly due partly to a fast-paced competitive environment with consumer choices, expectations and preferences which are always rapidly changing. Also, a digital textile print designer is taught through a recognized progression of primary, secondary and tertiary education; unlike the cave artists, they do not learn alongside a practising designer for all of these stages, and the large format inkjet printer is not sitting beside them throughout that period either. So, what the digital textile print designer learns of the tacit knowledge of other designers of digital textile prints is not the same kind of transferred knowledge as a cave painter would have experienced from other cave painters. One thing the cave painters did not have is the benefit of the additional knowledge that is contained in the body of archived artefacts, including documented and recorded information, plus the influences of the variety of social and cultural contexts in which digital textile print designers are now working.

This additional knowledge is contained in a range of media and, as McLuhan (1964) points out, the medium itself is the message so all versions provide different forms of communicated information. However, when people attempt trial and error alongside a skilled practitioner they are more likely to be enlightened than if they conduct trial and error sitting alone; so it is reasonable to assume it will take far longer for someone to learn by his or her mistakes when there is no one else around to watch and give feedback.

Another consideration is the act of making and the procedural knowledge that is preserved in a digitally printed textile. Even with one variable, such as the fiber content of the base cloth, there can be many different final aesthetics and products. The same is the case for different dyes, printers and images, meaning that the range of outcomes is extensive; however, the fact that one digital print can be unpacked to reveal so many different types of knowledge helps us to understand the range of complexities associated with designing for digital textile printing. This also provides us with a key factor in contemporary art and design practice, the concept of knowledge created through construction (Barfield and Quinn 2004), and a digital textile print is viewed as an object that has been made from an image before it is thought of as a product of a creative person executing his or her unique idée.

The quantity of knowledge tied up in contemporary designs and artefacts requires, according to Gardner (2006), a system that is capable of storing and organizing more information than our memories alone are now capable of handling. What systems must these ideas and images, plus this knowledge, be organized into? While some prehistoric cave painters had the opportunity to master in relative terms their domain, digital textile print designers are not able to be skilled at every single component of their craft, but they can, however, use short-cuts and mimicking tolls to emulate certain facets of those skills, and the overall essence of digital textile printing is what they can master. As Gardner (2006) advocates, creativity involves three characteristics: the practitioner, the domain with its acknowledgment of experience they are working in, and the mechanism that recognizes the standards of quality and judgments. So, while these three define the value of a digitally printed textile, for the prehistoric cave painter, only the first of these would have applied.

Throughout the history of applying an image created in one medium onto the surface of another, each development has evolved with distinguishable characteristics through the relationship between the materials of the image, the surface, the applying principle, tools and the synthesis through the emotional and sensational response of the artist to his or her initial idea or thought processes. These are the requirements for the print and within this group or set of materials, tools and creativity, the evolution from cave to digital textile printing can be charted and explained as a series of knowledge-gaining logical progressions that links the skills, creative impulses, responses to make with our own hands with and an inquisitive exploration and manipulation of the specifics of our surrounding environment. Banks et al. (1989) state that the essence of a culture is not within the artefacts themselves, but is evidenced by how society accesses, perceives and experiences them.

TYPES OF KNOWLEDGE IN DIGITAL TEXTILE PRINTING

The advanced technology used for digital textile printing encompasses the skills of previous experts whose accumulated knowledge developed the advanced technology at the core of this book. Much of the skill possessed by these earlier experts is now embedded as tacit knowledge within the new systems and is accessed by practitioners each time they engage with digital printing. By making explicit this tacit knowledge, and combining it with their own skills, the total knowledge displayed by practitioners is therefore greater than the intrinsic knowledge contained in the technology alone. The outputs of practice from less experienced designers, as it is supplemented with this additional knowledge, can thus be termed as skilled. As skill needs to be taught and developed, once practitioners learn how to embrace advanced technology, their outputs will demonstrate an even higher level of skill. The knowledge embedded in advanced technology thus enables designers to possess a higher skill level than would be the case if they were not using advanced technology.

BOUNDARIES AND RANGE

For those working in the area of digital textile printing, because there are so many new processes, techniques and materials, these can often obscure established beliefs of where the boundaries of the discipline lie (Carden 2011b). In order to locate an authentic digital print, its characteristics and properties, it is first necessary to define what it is and how best to describe it.

A digital textile print extends from a designer's worked out image stored as a digital file on a computer to the process of depositing droplets of dye from the print heads of a digital printer in the configuration of the digital image. In the case of reactive dyes, this image will require to be steamed to fix the image to the base cloth. While there are numerous techniques and processes involved in digital printing, none of them is unique to the printing of textiles (Hopper 2010); and because the computer is only a tool in the design process (Braddock Clarke and O'Mahony 2007), it is the practitioner who supplies the creative ideas. However, increasingly the technology is supplying designers with elaborate, satisfying and aesthetically challenging ways of working.

CONFIGURATION

In digital textile printing an image is printed on a base cloth. The image can be created in many ways, but the final version of it needs to be configured as a digital file. The digital file is then sent to the printer. The inkjet printer with independent print heads prints the digital image onto a base fabric.

To produce multiple runs of the original printed image requires a series of processes, technologies and materials to come together. Each aspect imparts its own characteristic on the final printed textile. This also means that the outcome is a conglomeration of many process, technologies and materials that represent the complex nature of digital printing.

REPRODUCTION

A reproduction is not the same as the original version of something. Benjamin (1936) explains that the key fact here is that the original of anything, such as a work of art, was produced at a particular point in time and all aspects of its creation were undertaken during that same period. Those who viewed the piece when it was first produced would be seeing it for the first time, would be unaware of how it would be reflected upon in the future or what consolidated ideas might be formed about it upon it over time. This leads us to question: who owns the original of a digital textile print, where is it located, and how can it be stored? No one print can claim to be any more authentic than another because the means of its production are contemporary to each print; and this is the same whether the digital image is printed today, next week or next year. What moves intact through time, though, is the combination of the digital file and the digital printing of this file; neither can produce the final print without the other, so are interdependent. The digital image exists only in the computer and the digital printer needs a file to print, so without each other neither can realize anything tangible.

What happens in the future if the technology currently used to interpret the original image as code, send it to an inkjet printer, and send the information to the print heads to deposit the dye in the formation of the image is superseded by more advanced technology? Where is the authentic version of the digital image in this scenario? Do the hardware and software used to create an original digital print ultimately mean that its existence is predetermined by the lifespan of this technology?

According to Rancière (2004) a number of factors reveal the relationship that exists between a culture in which an original design or artefact was created, and the aesthetic background that existed at that time. And Benjamin (1936) believes the original design occupies a unique presence in relation to this marker in space and time, so the nature of this presence and how it can be transported over a distance of months or years are key to determining what makes it authentic.

MASS PRODUCTION

There is a significant difference between the processes used to produce an artefact and the act of reproducing it. Making a copy of something is a very different process from creating an artefact in the first place. Practitioners use numerous decisions and judgments when they design something, including explicit, tacit and procedural types of knowledge; however, the same skills are not needed for the reproduction of the artefact. An original print can be more difficult to pin down than, for example, a painting. Firstly, the original design for a print would be created from the outset and with a reproducing technique in mind; secondly, when multiple prints are produced, it can be difficult to determine which one of them was produced first, and deciding where the authentic version of the print exists is even more problematic.

Looking for the authentic print is further complicated in a digital rather than an analog situation. The means of reproducing a traditional photograph, using equipment such as a non-digital camera, is quite different from the process required for a digital print, in that it uses a single, physical, non-digital negative. However, with a digital print the process of reproducing involves crossing from a digital state to a non-digital one, from a digital image to a material interpretation of that image on a base cloth, and the final artefact has a markedly different aesthetic from that of the original light-sourced version on a computer monitor.

The idea present in a designer's mind at the start of the creative process influences the selection he or she ultimately makes regarding which base cloth to choose; this selection in turn impacts on the final look of a printed textile in terms of its hue, tactile quality and color. The fabric also dictates the function to which a textile is suited, as well as its intended market. This demonstrates why there are multiple variables to take into consideration when planning how each outcome of digital textile printing will appear.

ORIGINAL CONDITION

Benjamin's (1936) position maintains that the authentic reflects the ability to possess characteristics of original physical condition that can also result in a change on ownership. In traditional photography, a negative enables multiple reproductions of an image to be produced, although it is not always possible to find a definitive original photograph. By comparing traditional photography with digital textile printing we see there are some similarities between the negative of a photograph and the file of a digital image, both represent a single instrument for reproducing and each is capable of being manipulated prior to multiple printings. However, while the negative of the photograph is a single physical item, the counterpart for digital textile printing comprises not one but many material and non-material components, processes and technologies. These range from the digital file to the droplets of dye that are distributed through the print heads of the inkjet printer towards the base cloth. The means of reproducing does not include the final printed artefact, so it would appear that the term digital textile print encompasses a number of processes, including everything that is necessary to print a digital image digitally, but does not include the dye-formed image once it is in a physical state. The final printed textile, similar to a traditional printed photograph on photographic paper, is a post-condition of the printing process.

PRECONDITIONS

Flusser (1983) considers that the world, and this includes the camera, is a precondition for a photographic image. This could be interpreted as meaning that digital landscape, such as the computer, is a precondition for what will become the digital textile print. Flusser suggests the negative of a traditional photograph is authentic, and this implies that its preconditions are outside the authentic; it also implies that the authentic represents the boundary of what is real in the photographic process. An image, says Merleau-Ponty (1960), regardless of how it is produced or reproduced, is not located in the viewer but is instead to be found outside the person. It is a thought woven to make a connection between two non-touching entities. This would imply that a digital image is only real when it becomes integrated into a similar woven set of ideas about what an image is, or is going to become, and that idea is situated somewhere between the digital file and the surface of the base cloth. But, for this to be achieved, the boundaries of a digital print must retain links to the image as it is when a digital file, and the image as it is when a dye-based print, traversing the non-material and the material. If, as Merleau-Ponty (2002) also claims, regarding creativity across diverse media, neither of these processes has comparable aesthetic qualities or attributes, then a system of equivalence must exist to connect them. Thus, for the digital print to be reproducible, it needs to have a viable constant, one that comprises the authentic traversing medium, the image as digital file and the image as digitally printed in material form at the point of output.

COMPONENTS AND CONDITIONS OF REPRODUCTION

For digital textile printing the components of the means of reproducing are the printer and the digital file. Although the printer requires something to print and the digital file does not normally exist within the printer, this implies that the means of reproducing necessitate the printer and the computer to be treated as interdependent. It thus seems appropriate to consider them as one entity for the process of digital textile printing. There are four factors that determine the characteristics of a digital textile print:

1 The visual aesthetic of the original image.
2 The process of production.
3 The mechanics of reproduction.
4 The base cloth upon which the image is printed.

Of these four factors, the first is a precondition, the second is a condition, and the third and fourth are post-conditions.

The authentic of something represents all that can be transmitted from the start of its existence in the world to what will become its history, that is, what can be conveyed from one point in time and space to another. The processes, technologies, material and non-material elements of digital textile printing when grouped together are capable of being moved from one time and location to another, though undoubtedly they will deteriorate according to the effects of time, but this fits with Benjamin's definition of the authenticity of an artefact (Carden 2011b).

CONCLUSIONS OF THE ESSENCE OF DIGITAL PRINTING

This chapter has shown that, when practitioners create artefacts using advanced technology, they are producing outputs that contain knowledge from three main sources: the information contained in the materials and technologies associated with older technologies; the skill or previous experts embedded as tacit knowledge in the advanced technology; and the mastery of the current practitioner. The creativity that led to our contemporary culture in which digital textile printing is located can therefore be traced back to when people began to form language and produce marks on cave walls over 10,000 years ago. The route taken and the influences that shaped it along the way will help us to better understand the factors of how today's digital printing will most likely shape its future direction, and some of the main areas it will influence.

To summarize: the digital image is represented as a file on a computer and the digital print is the image as file combined with the image as dye. An inkjet printer with independent print heads prints the digital file. The process for digital printing necessitates the print action to be joined with the digital file. All of these processes and technologies, non-digital and digital components, are necessary for the digital print to exist. Although the coded file informs the print heads of the printer and the print heads need this information in order to act, so the file and the print heads respond in a manner that is similar to the brain and the hand, but once the dye has been deposited by the digital printer and been fixed onto a base cloth, the authenticity is lost and the digitally printed textile is one of many. This means that the areas in which the digital print can be altered and manipulated are outside the boundaries of what is considered to be the authentic digital textile print.

7

INVESTIGATING DIGITAL TEXTILE PRINTING

This chapter looks at the opportunities that exist for investigating multiple perspectives of digital printing, by placing it in a research context. Due to the nature of the printing process and the variety of materials that can be used as substrates, advanced technology is increasing the breadth and scope of what designers are able to incorporate within their work. As a consequence of these new opportunities, practitioners are also increasingly positioning themselves as researchers in their own studio inquiry. This means designers can consider the implications of how they create with digital textile printing, while simultaneously allowing them to use previously mastered skills, and reflect on alternative ways of making. This chapter also explains why situating advanced technology as the primary focus of research informed art and design practice can inspire designers to reflect on how they approach studio practice, and better understand the impact digital technology has on the aesthetic of final artefacts. Using philosophical and sociological positions such as the models of Heidegger (1977) and Eco (1989) helps us explore why the non-human relationship practitioners have with machines leads them to instinctively combine attractive features with functional aspects to avoid feeing enslaved by technology. This leads us to speculate that practitioners are increasingly encouraging technology to act for them in a manner that complements their person way of working. However, Gardner (2006) maintains practitioners must first master skill if they are to realize its full potential, so although advanced technology embodies a certain amount of skill from previous experts, such as those who originally developed the technology, a machine cannot master additional skills, it only outputs what it was designed to do. Therefore, if artefacts are to have increased depth, they must demonstrate a level of skill that only practitioners can provide. Although advanced technology enables practitioners to create work increasingly cheaply, quickly and easily, this can only be carried out successfully when those using technology have previously mastered the skills behind it. This works in two ways. First practitioners are less inclined to feel intimidated by technology if they are already experts in whatever areas the technology is aligned to, and second, when practitioners have sufficient experience of other, older technologies. This supports them to knowledgeably include additional processes and systems within their practice, thereby increasing the potential benefits of the new technology.

ORIGINS OF DIGITAL TEXTILE PRINTING REQUIREMENT

Digital was originally developed to provide more efficient forms of analog processes, but its ability to mimic was perceived by many practitioners in this study to be fake and their personal preferences were still with non-digital systems. This impacted on the decisions they made in the studio. The more digital was seen to copy non-digital the less value was attributed to it by an expert audience. This led to a digital versus non-digital dilemma for many of the practitioners. While combining digital and non-digital could produce exceptional results, there were still clear advantages to be had, individually, from each. In an attempt to utilize the best qualities of each, practitioners found that when they skilfully joined them together, the results provided greater creative opportunities and novel outcomes than were possible with either process alone.

As digital continues to evolve, older technologies and systems are disappearing, either due to cost or reduced demand. The features and characteristics of the traditional processes and technologies that do survive are being redefined as luxury or unique. Advanced technology is increasingly echoing human specialism and, as a result, demands on practitioners are changing. This study has shown that artists and designers are altering the way they practice in response to the disappearance of older systems, while at the same time, embracing new developments in technology. Practitioners are now making advanced technology work for them in a reflective manner, in a way that complements their individual preferences and personalities. It is the humanizing of technology, making it work in ways that it was not originally designed to do, that affords the most efficient use of advanced technology. This project has found that practitioners are not content to design for technology, but rather, are asking technology what it can do for them. By realizing their digital demands they are helping hardware and software developers to improve existing systems and also provide future artists and designers with a guide to conducting reflective studio practice, thereby enabling practitioners to enhance their creativity.

THEORY

Friedman (2008) points out that if we are to fully comprehend how design can operate within the context of different types of processes, media and information it is first necessary to develop a theory. However, a formal theory is developed using significantly different conceptualizing tools than those generally found in the studio environment, so by using design practice as a way of generating data to reveal an in-depth, personal view of working with advanced technology can be revealed. This approach is adopted from a model used extensively in social science for theory building (Glaser and Strauss 1967). Having first focused on studio practice, it is then possible to explore the phenomenon in relation to other practitioners working in neighboring fields, in order to build a theory that can help a wide range of practitioners understand how they might more effectively create with digital technology. Building theory through digital textile design practice empowers practitioners to explore the importance and application of theory in the context of advanced technology. By focusing on this gap in knowledge an improved understanding of the application of digital technology, both its benefits and disadvantages, can be revealed so that artists and designers may be supported to select combinations that work better for them, thus enabling practitioners to take maximum advantage of the creative options available to them. However, until recently a theory for digital printing had not been constructed (Carden 2013b).

THEORY BUILDING

Through a practice-led and reflective approach to digital textile printing a number of ways can be identified in which craft and digital processes are capable of being combined to produce innovative new techniques. By using textile practice, processes can be developed and analyzed through studio inquiry to build a theory in craft and design research. Also, by investigating digital textile printing from a variety of perspectives, alternative ways may be identified in which advanced technology links textile design practice to other design domains, and these connections enable ideas to be generated that are then synthesized to form hypotheses.

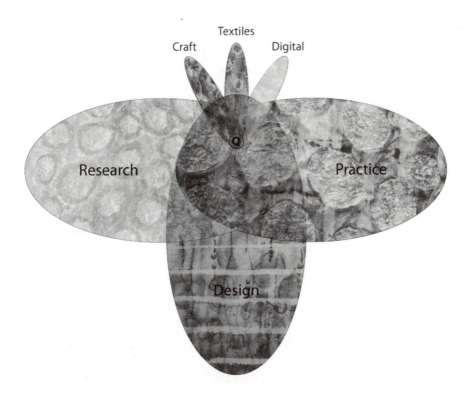

Figure 7 Digital textile printing as research. Susan Carden 2012.

POSITION OF DESIGNER RESEARCHERS

While designers are not impartial spectators of their own practice, they can find that by reflecting on the various documented outcomes their role as designer allows them to gain unique perspectives on practice. They can observe and experience it from the point of view of someone on the inside who makes, rather than solely witnesses, records or surveys, to allow associations and theoretical assumptions to be made. As Gray and Malins (2004) point out, the use of creative practice is a subjective process, and there may not be any clear universal truth to be had anyway. Scrivener (2002b) believes that Schön's (1983) definition of the reflective practitioner enables us to access the way creativity works from the inside, including influences from previous experiences, by enabling multiple perspectives of the act of making to be revealed. Scrivener and Chapman (2004) also propose that this reflective practice is grounded in current work, and can be further realized through future projects. This perspective means that an interactive cycle carries reflective practice forward from an initial phase, through consecutive stages in which the various aspects of the phenomenon under investigation are repeatedly revisited. The process of reflection provides designers with information from three main areas: before, during and after the project, and supplies them with extensive documentation on the work, details of the outcomes and the decisions that were made. Niedderer (2009d) explains that knowledge generated during the making process, procedural knowledge, is partially communicated through verbal means or by providing others with detailed descriptions. However, she maintains that much of this information can only ever be transferred or made explicit by demonstrating either in person or by the use of video recordings, as explored by Philpott (2011); this enables someone other than the practitioner to learn how work was created.

PRACTICE-BASED AND PRACTICE-LED

The terms practice-based and practice-led are routinely used to describe research undertaken by practitioners in the areas of art and design. As Biggs (2002) explains, practice-based research includes the means of making the artefact at the center of a study while Candy (2006) understands that practice-based research involves the artefact itself becoming the focus of the investigation. These definitions lead to the conclusion that practice-based research should focus on the artefact, or how it was produced, in order to generate new knowledge. Practice-led research must lead the researcher (AHRC 2007) and is therefore more concerned with the undertaking of practice and how the activity is able to lead to new territory that will provide data for the researcher to analyze and synthesize. Sullivan (2005) suggests that a challenge for contemporary researchers is to alter the types of investigations they conduct to include studio practice.

PRACTICE-BASED RESEARCH

One of the main reasons why research studies in the field of design frequently fail is due to their inability to interact with grounded theory, with the result that projects often do not develop theory out of practice (Friedman 2008). It is only by reflecting on practice that explicit knowledge will emerge and through interaction with grounded theory that practice can yield theories (Glaser and Strauss 1967). Understanding and explaining what designers do during their studio practice can be assisted by the common conventions of perception, language and methods of articulation (Niedderer and Roworth-Stokes 2009). Through these interactions, peer responses provide useful, albeit subjective, contrasting views of design practice involving the digital reproduction of images. Also, naturalistic research, although taking place in several different studio settings, is only as real as the practitioners themselves are willing to reveal (Erlandson et al. 1993).

ONTOLOGICAL POSITION

Heidegger (1962) highlights the importance of understanding the truth surrounding our existence in the world and he makes a clear distinction between the terms ontic and ontological; ontic relates to entities that are real, rather than phenomenal, while ontological deals with the characteristics or nature of the objects being investigated. The ontological position of practice leads to a number of assumptions: to begin with, it can be described as a materialist ontology (Willig 2008). An ontological position can be either realist or relativist, with a realist ontology comprising physical objects and factors that impact on each other, so materialism is included here; whereas, relativist ontology disagrees with the idea of orderly relationships existing between objects (Gray and Malins 2004; Langdridge 2007; Polanyi 1974; Willig 2008).

POSITION OF MATERIALS IN RESEARCH

While Nimkulrat's (2009a) study into her artistic practice was investigated from a key position of material in relation to technical skill, when advanced technology is included in this equation, Hohl (2009) suggests the making hierarchy is reconfigured with the idea or vision placed ahead of material and skill. This premise concurs with Harrod's (2007) view that advanced technology provides the practitioner with additional time to think and contemplate due to the number of labor-intensive tasks that have been eliminated from the making process. Margolin (2010) highlighted that designing creates new products and so any investigation into design should, to a certain extent, also focus on how we engage with the design process. However, concerning the design process, knowledge will only be of maximum value to design researchers when it is open and transparent. As Friedman (2008) explains, we need to be explicit in our communication if theories are to be developed and shared, but he warns that some designers mistakenly confuse practice with research, pursuing their normal design activities rather than using inductive inquiry to develop theories.

TRADITIONAL SKILLS

Nicol (2009) suggests that few contemporary designers possess the traditional skills and practical knowledge essential in her specialist field, meaning that when practitioners attempt to combine traditional practices with new technologies, their lack of expertise often results in the balance shifting towards advanced technology. Althusser (2008) advocates that to produce aesthetic value from the visual arts, we must spend considerable time developing the skills that are required, and we should not be too quick to move on to the next challenge, otherwise we may not fully acquire the knowledge we have been looking for, merely an ideology of it. Furthermore, Weiss (2008) maintains that if care is not taken, the process of mastering craft skills could be overlooked, resulting in artefacts that lack the qualities that often make them distinctive.

However, Bolt (2006) argues that artists and designers are fortunate to be in a unique position in which making with tools and materials takes place in the naturalistic environment of everyday life. Bunnell (2001) recalls that during her PhD viva voce the examiners repeatedly raised concerns about the unique nature of her single case study, questioning its transferability to the wider realms of practice. Valentine (2011) suggests that although materials and technical skills are important, her research found that the idea and the concept are the main drivers for craft, and Bunnell (1998) was able to demonstrate that she had disseminated the tangible outcomes of her vision and concept to a wider audience through exhibiting her work. As technology increasingly becomes more familiar to artists and designers, as well as the audience perceiving the artworks, this could result in ideas and concepts rather than the advanced technology, driving creative practice (Hohl 2006). Gardner (2006), Niedderer (2009a), Polanyi (1974) and Schön (1987), identify the three primary types of knowledge contained in outputs of digital textile printing. When artists and designers work with digital technology the combined skill from each of these components enables them to create more efficiently and produce exceptional outcomes.

SKILLS IN PRACTICE-BASED RESEARCH

Nimkulrat (2007) believes craft-based art and design discourse is structured around the acquisition of technical skills. She maintains skills are taught first and then materials selected according to their appropriateness to those skills. The usual hierarchy, according to Nimkulrat (2009a), begins with skills, although adding a further component advances technology and the connections formed through its application. Nimkulrat's (2009b) study challenges the existing order by placing materials at the top, thereby inspiring artists and designers to reconsider how they approach their practice from a materials-based perspective. However, we can suggest advanced technology be given this primary position, so that the impact of new technology can be more clearly explained within the context of textile design practice.

CRAFT SKILLS

While craft allows practitioners to absorb ideas and aspects of expertise from many different areas of knowledge, the development and value of procedural knowledge found in craft is also allowing makers and researchers to extend their influence from practice-based contexts to new areas of processes and theories. These emerging opportunities are complementing and expanding the practice of contemporary craft-makers, providing handcrafted artefacts with both a new audience and relevance. The final outputs are thus enriched through associations with a wider range of perspectives. The capacity to forge new relationships through the act of forming artefacts can be analyzed from multiple positions, including material, function, process, aesthetic and concept, all of which inform the integrity of the crafted object. While materials suggest patterns, structure, composition and form (Yanagi 1974), and every material has its own language (Dewey 1934), it is understandable that crafting with fabrics and dyes will result in outcomes that are informed as much by the practitioner as their materials. When artists and designers extend the range of tools, materials, techniques and processes at their disposal, to include craft and digital technology, it is not surprising that the resultant dialogue creates new connections and the synergy of these opportunities can produce extraordinary outcomes. Each component, skill, material or technology, is so intertwined within digital textile printing that a change in one necessarily impacts on all the others. Therefore, as more digital prints are produced, practitioners must identify new ways of embracing potential increases in output while simultaneously making digital prints more, not less, distinctive.

PRACTICAL EXPERIMENTATION

Carden (2013b) started to experiment in her studio by considering the individual elements that come together to form digital textile printing. She looked at the boundaries of the domains and the materials and processes required to produce digital textile prints. She was especially interested in the large format inkjet printer, and decided to limit herself to machines that print with reactive or acid dyes on cotton, silk, wool and viscose. As her practice mostly involves natural fibers and dyes, she decided to use the same materials in craft-based experiments. Reactive dyes are combined with an alginate in solution to help them flow through the print heads, and alginate is also used in a coating for the base cloth, comprising alginate, urea, sodium carbonate and water. After printing, the fabric is steamed, fixing the reactive dye to the surface of the substrate so she realized that alginate and water are the common materials used in the digital textile printing process. She began by exploring the potential for these two ingredients to be used as a basis for new techniques in digital textile printing.

EXPERIMENTATION WITH TRIAL AND ERROR

At first Carden began to play with a thick alginate gel to see if it could be used in any other ways, such as applying it to the digitally printed base cloths. She experimented with substrates that had already been digitally printed upon, such as light silk chiffon, medium weight silk satin, and heavy silk dupion, as well as different weights and qualities of paper. She discovered that by combining a glossy paper with the alginate solution had no effect, but a digital image on rough paper was mobilized by the mixture and, when she placed some silk satin on top of the saturated paper, a large proportion of the image transferred from the paper onto the silk. When she rinsed the silk and gently washed it in soapy water, the transferred image stayed fixed to the silk. The outcome was unexpected as she had not pre-coated the silk before trying this experiment, and she did not need to steam the silk afterwards to fix the image. The outcome of this was that two of the processes that use a great deal of energy and chemicals in digital textile printing were not required.

She saturated the paper image for over half an hour and noticed that a range of colors had moved to the silk and some of the dye had actually adhered to the silk. She observed that certain colors of dye or ink invariably traveled to the silk, while others stayed on the paper. It was the black dye that would not transfer, and she then wondered if it was perhaps the type of dye or ink that was causing this phenomenon to occur. She looked into the makes and types of ink that are available for desktop printers: for example, an HP Designjet 10ps and a Canon MP190 use dye-based inks for their color cartridges but have pigment in the black ones. As it was the black that did not transfer from either printer, she checked the cartridges of the ones she had used and found that these were indeed pigment inks. However, she later found that a black from the large format inkjet printer, a Stork Mimaki TX2, was transferrable and it turned out to be reactive dye, so this confirmed her initial reasoning.

When practitioners hand dye fibers or fabrics, they routinely use vinegar as a natural fixative, so she tried different types of vinegar with the transferred prints, including vinegar from cider, sherry and white wine, to see if these would have any effect. She found that it was possible to wash the transferred silk samples at quite high temperatures, up to 40 degrees, without any colors running or fading. However, most of these vinegars also tinted the base silks, and only the white wine vinegar did not stain the printed textiles, so she sought out other potential liquids that could provide a similar fixation without altering the colors that had already been achieved. Vinegar has a pH value of 3, so she looked for something with an even higher acidity level, such as fresh lime or lemon juice. She discovered that lemon juice worked remarkably well and the transferred image stayed clear and vibrant. The final outcome was not only well fixed, but appeared to be better integrated within the fibers of the base silk than the original digital textile prints had been from a large format inkjet printer, and the image appeared to be penetrating further into the base cloth. This helped remove the impression that the image was skimming on the surface of the silk. She then experimented with, and checked the pH values of, a range of substances and noticed that the further a substance is away from a neutral pH of 7, the better its ability to transfer and fix an image from a Hewlett Packard or Canon desktop printer via paper onto a base cloth such as silk.

By further developing of these techniques, involving a significant amount of experimentation, play and error, in the studio, she was able to discover a series of unusual craft-informed processes. She describes these as interventions because they are situated and in fact grounded in the application of craft within the boundaries of digital image generation and printing.

CRAFT TECHNIQUE I (SEE PLATE 24)

Inverse batik. Take an un-coated substrate; paint a pattern using the pre-printing alginate solution then digitally print with the large format inkjet printer. The painted area holds the digital image, while the un-painted sections of dye wash off, leaving a batik-effect. This technique is similar to traditional wax batik; however, whereas wax resists the dye, in this technique, the alginate solution holds onto the dye.

REFLECTION ON CRAFT TECHNIQUE I

"The image is well wrapped around each thread, giving a good, overall coating. There is an even covering on the surface of the substrate, nearly as much on the rear. The dye almost seems to be incorporated into the yarns as clear light is bouncing off the surface and seems to be skimming over the top of the image that is bonded with the substrate. Very much seems to be part of the structure of the fabric. Also, at the sides of the image there are no hard edges, it is as if the image floats a bit along the warp and weft ends, merging a little into the surrounding fabric, blending at the perimeter and diffusing around the sides of the printed area" (Carden 2013b).

CRAFT TECHNIQUE 2 (SEE PLATE 25)

Raised print heads on hand-manipulated substrate. First, hand-pleat the base cloth, then raise the print heads on the large format inkjet printer up to 10 mm, then print. The final image is diffused and the handcrafted nature of the manipulated substrate produces a randomly distorted image.

REFLECTION ON CRAFT TECHNIQUE 2

"The digital print is obviously on the surface, with the transferred image slightly beneath and penetrating through much more to the rear. The black component of the original digital image has somehow moved into the core of the substrate and is visible on the back which it wasn't before the transfer process was applied, although the rest of the digitally printed colors are not. It is strange that the digital-black moves further back, as black is the one color that does not normally become mobile in the transfer process. The transferred image merges with the overall substrate filling in the areas that were not fully coated before with the digital print alone. The digitally printed image slightly sits on the transfer image bur, because the latter is well integrated with the substrate, the sides of each yarn are coated in color, so it has a comfortable knitted together, blended look. The transfer breaks up the digital image a bit because of this coating of the un-dyed fragments of the substrate, and the structure is far better integrated into the final combined set of images as a result" (ibid.).

CRAFT TECHNIQUE 3 (SEE PLATE 26)

Transfer image with inverse batik. Transfer a digital image, paint a pattern on the textile with the pre-printing alginate solution then digitally print using the large format inkjet printer. The final multilayered image is a combination of different dyes, incorporating an inverse-batik effect and multiple digital prints.

REFLECTION ON CRAFT TECHNIQUE 3

"The transferred image is slightly more clear on the surface, but quite obviously penetrating through to the back. The batik area has blocked the transferred image with fairly solid edges. The structure of the substrate is well integrated with the overall transfer image. The light is reflected in all directions along the lines of the weave's structure, and this helps disperse the light with an interesting roughness, especially along any slubs in the fabric of the substrate, which are highlighted towards the light source" (ibid.).

CRAFT TECHNIQUE 4 (SEE PLATE 27)

Batik transferred image. Apply melted wax as a pattern on the substrate, transfer an image and then iron the wax between two sheets of paper to remove it. The final textile is an example of a batik informed transferred digital print.

REFLECTION ON CRAFT TECHNIQUE 4

"Various levels of dye saturation occur. The transferred image is seen on the surface and rear, and is solid around the batik areas. The digital image that was previously printed sits on top inside these subsequently batik areas, but is seen on the surface only. The transferred areas are less prominent on the back but are quite strong and dispersed with unprinted segments where the transfer process has detached itself during the transfer. The layering of images is reflected in the structure of the substrate, the transferred image is slightly more integrated with the yarns than the digitally printed image, which sits rather differently on the warp than it does on the weft. On the areas where both are present, the transferred one goes right through to the rear with the digitally printed one sitting slightly on top of it. The transfer helps to blend the digital with the natural color of the substrate, while the batik is clear and the digital image sits on top within the waxed area" (ibid.).

CRAFT TECHNIQUE 5 (SEE PLATE 28)

Stencil transferred image. Cut out a stencil from waxed paper, lay on top of the substrate, and apply transfer process. The final piece is a stencil-effect transferred digital print, formed from the original scalpel-crafted stencil.

REFLECTION ON CRAFT TECHNIQUE 5

"The horizontal slubs of yarn absorb the greatest quantity of dye, and this separates close to the colored areas. The darker dye travels to the outer edges of the dyed areas while the denser regions randomly pick up the dye on the horizontal yarns more than they do on the warp ends. There is a clearly defined area of separation between the sharper edges and the sides of the stencil but from a distance the broken regions give a watercolor appearance that although appearing to be random on the surface, as though it was not fully absorbed on the surface, manages to penetrate right through to the rear more than any digital print would from a large format inkjet printer; so it looks random and slightly broken up while, in fact, it is fully absorbed by the fabric and yet still looks sufficiently random to be a one-off watercolor effect stencil" (ibid.).

CRAFT TECHNIQUE 6 (SEE PLATE 29)

Multilayered transfer image. Apply melted wax, transfer image, remove wax; repeat with a different image. The outcome is a textile patterned with multiple-transferred images, for example, created from a transfer of a photograph of an original glass mosaic, and the process repeated using the same glass piece after it has been fused in a kiln.

REFLECTION ON CRAFT TECHNIQUE 6

"The colors are distinct and separate from each other, and the image is clearly composed of a number of individual colors. There is evidence of a slight overlap of dye, when looked closely under a magnifying glass, and a very slight un-dyed edge opposite this, giving an attractive, random unevenness. Each overlap contains at least two layers of dye so these areas are darker, giving extra depth at the sides of some blocks of color. This means the image is not visually flat, but has almost a three-dimensional effect at the sides of the areas where overlapping occurs. The dye is clearly seen on the rear, not as strong as on the front, but even the overlap is visible. The colors are more solid, as these areas are clunky. The overlaps are a combination of all the colors from each side adjoining it, so the character of the overlap is determined by the dye, so comprise those colors over-layered or combined. The dye soaks in well into the structure of the substrate, and the un-dyed edges of the image contrast distinctly with the dark, repeatedly printed edges opposite. If it is dark on one side of the printed shape, the other side will be unprinted to reflect this displacement. The movement-in-action of the screen-printing process is unique to each printing session" (ibid.).

By closely observing and comparing each of the printed textile samples, Carden produced a diagram that demonstrated how the digitally printed images break up on the surface of different substrates. These examples demonstrate that the luminosity of each digital textile print differs depending on the base cloth, process or technique that is used.

The effect that is visually unexpected and unpredictable occurs when there is an uneven emission of light. This occurs when the surface of a fabric is visible through the dye, and when the dye

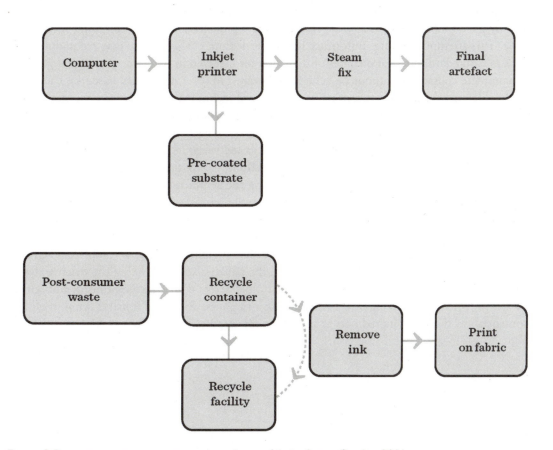

Figure 8 Post-consumer waste: paper to print on fabric. Susan Carden 2011.

is absorbed into the fibers in such a way that it is altered by the physical structure of the cloth. Therefore, the fabric causes the image to look the way it does, and the image causes the final digital textile print to look the way it does. This mutual state of causality results in textiles that are concatenated, that is to say, they are so tightly bound together as to be interdependent, and are therefore evenly balanced. What this means is that within the limits of digital textile printing it is possible to produce different types of surface color on the yarns and this creates a variety of states of luminosity. By combining these alternative outcomes a practitioner can produce textiles with multiple aesthetics, making the digital technology respond to the ideas and materials available to the designer. Rather than let the technology dictate the limitations of what a practitioner can achieve, this allows the designer to work with the technology to make it to work more efficiently for them.

INFLUENCING FACTORS

The key factors of digital textile printing outlined by Campbell (2008) are significant, even though his focus is on apparel design. Campbell cites, for example, the importance of the height of the print heads to ensure accurate printing, but it is through this facility that it is possible to create non-uniform outcomes. He also points out the importance of ink placement to make sure the printed image on a base cloth is as close to the visual representation on the computer monitor as possible. This attempt to create uniformity is another feature challenged by new techniques. The natural, non-standard, physical qualities of the materials involved in digital or non-digital printing mean complete consistency cannot be achieved anyway. However, the issues raised by Campbell show the dilemma for designers working with advanced technology, such as the goal of accurate, uniform, predictable printed outcomes, required for mass production, juxtaposed with the aim of bespoke, slow designs for practitioners who often have other agendas.

While digital technology is able to foster several varied roles, including collaborative practical experimentation and process forming through connections, within textile design practice this is still largely uncharted territory. It is by cyclically revisiting textile practice, and reflecting on the work created, that a unique perspective on how a digital textile print designer works and the outcomes they produce will be revealed.

Reactive dyes are combined with alginate in solution to facilitate their flow through the print heads. The substrate or base cloth is coated with a different alginate solution, comprising alginate, urea, sodium carbonate and water. After printing, the fabric is steamed, fixing the reactive dye to the surface of the substrate. The common materials used across digital textile printing are alginate and water so, when investigating how new techniques might be developed it makes sense to explore the potential for these two ingredients as an avenue for creating innovative processes.

8
CROSSING DISCIPLINES

This chapter reviews the outcomes of investigations into how practitioners working in disciplines that border digital textile printing feel about the opportunities they are experiencing. With digital printing a common process across disciplines, artists and designers from different studio backgrounds are approaching advanced technology in similar ways. This chapter looks at the relationships that exist between digital printing and handcrafting across a range of disciplines, such as textile design, print-making, typography, illustration, filmmaking and photography. It seeks to

Neighboring disciplines that use non-digital and digital technology

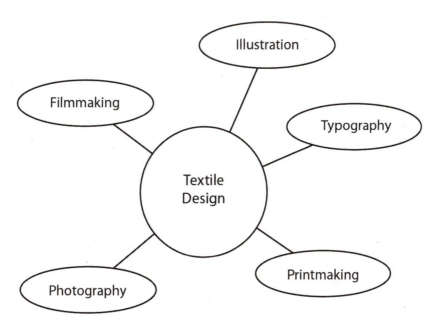

Figure 9 Textile design and neighboring disciplines. Susan Carden 2011.

define how a variety of practitioners are able to use digital printing and highlight the challenges designers are facing. While digital printing offers many advantages, there are a number of features that some designers are less willing to use. Through a number of interviews with expert practitioners from disciplines that include and neighbor textile design, key elements for, and against, analog and digital processes are identified, explained and analyzed.

DIFFERENT DISCIPLINES OF DIGITAL PRINTING PRACTICE

Recent research involved conducting a questionnaire survey with practising artists and designers (Carden 2013b). Participants were selected non-randomly based on the studies of Hall and Hall (1996), and Oppenheim (1992), who suggest using an extensive sample in order to gather a wide representation of experiences and opinions. Those questioned were practitioners in textile design plus the neighboring disciplines of illustration, typography, photography, filmmaking and print-making (Scagnetti 2012). Participants were experienced practitioners with skills mastered over many years and represented an equal number of male and female interviewees, in order to achieve a gender balance. The artists and designers contacted were all from the U.K. because the tools, facilities, institutions and levels of access to advanced technology for these practitioners would be more equitable than if they had been selected from far-flung geographic locations. The questions were based on Spradley's (1979) model that includes descriptive, structural, contextual and evaluative questions, and the interviewee was mindful of cross-disciplinary variations in terms of meaning and terminology. For example, in textile design the substrate is normally a form of fabric, whereas in typography it is generally a type of paper. The questions were deliberately neutral rather than value-laden (Langdridge 2007) to encourage the participants to freely express themselves.

As well as a questionnaire, semi-structured interviews were conducted as a method that provided purposeful conversations (Robson 1993) aimed at revealing the meaning of the use of digital and non-digital processes within art and design practice. Gray and Malins (2004), and Keats (2000), suggest interviews are a valuable way of eliciting in-depth responses and views, so the outcome of these provided interesting insights into the creative process.

DIFFERENT PERSPECTIVES

Advances in technology are influencing how digital textile print designs engage with skill and expertise. By exploring what digital textile printing means, why it is created the way it is, how it can be approached from different perspectives, the field is increasingly providing new ways for researchers to investigate its potential and, as a result, create new knowledge. On one hand it can be viewed in terms of methods of production, the materials and processes and technologies that are used; on the other, there is the perspective of the audience (Heidegger 1962). By taking this approach, the researchers are placing themselves in the position of the artefact (Merleau-Ponty 2002), and this can be further helped by the investigator making a version of the object with their own hands, using tools that may include digital technology, in order to better understand the piece. It is essential, claims Dormer (1997), for a researcher to also remember that the process, such as advanced technology, is the culmination of many previous tools, so that the ideas and concepts of its development may be unpacked and used to build on existing knowledge.

TRANSFERRABLE COMPONENTS OF DIGITAL TEXTILE PRINTING

There are a number of key components of the digital textile printing process that are significantly different in a non-digital context than they are in a digital one. A result of this is that the act of creating for digital textile printing is quite unlike traditional older methods because it calls upon a variety of different senses, instinctive responses, techniques and materials, all of which lead to an alternative distinctive design approach. The components in the context of digital textile printing can be divided into two categories: the physical properties of the materials and hardware that are used, and distorting, manipulating properties that impact on those physical features.

Physical Properties

The opportunities available to a designer through advanced technology enable them to approach their practice in many different ways, for example image manipulation and multiple layering of narratives. This means designers use the components of digital textile printing in such a manner as to make the advanced technology work as naturally as possible. This can be about manipulating the digital technology in an attempt to push what it is capable of providing beyond the point where a coded file is sent over to a digital printer for output. In traditional, older printing practices the human involvement is carried on throughout the printing process, from image creation and dye selection to the hand printing of the final motif onto fabric, whereas for digital textile printing the personal hands-on involvement ceases when the coded image or file is passed over to the machine for printing (Carden 2013b).

Digital textile printing provides a designer with millions of color choices. By simply looking at a printed fabric a person can usually tell whether a large number of colors have been used or not. If multiple colors were applied, then that textile is generally digitally printed, rather than being produced by screen-printing (Carden 2013b). Even if the print has undergone post-printing treatments, such as distressing in an attempt to make it appear to have been produced prior to the advent of digital printing, ultimately it is the large number of colors used that can confirm if it is a digital print or not.

In a screen-print each color is applied separately; it is distinct and therefore produced at a different time (Carden 2013b). Because each color pass is performed independently the combined outcome can result in overlapping layers of dye that have built up around the edges of individual separations, and these form slight raised areas. Also, occasionally, unprinted small areas are located where the dye has been omitted altogether due to the screens' misalignment. Any unevenness on the surface of the fabric, such as this build-up of dye, can create an interesting physical phenomenon that is indicative of a screen-printed textile and this can vary further when printed by hand. In contrast to this, a digital printer applies dye in one pass and therefore has no such overlapping so it is usually flatter and smoother.

Although a textile designer can use unlimited colors and create infinitely complex images, there is only one type of dye associated with each digital printer, and this is predetermined at the outset of the design process (Carden 2013b). The pressure of dye forced through the nozzles and the height of the print heads are also fixed prior to printing, so the designer is restricted during printing in these component areas. However, even though there is an almost infinite range of colors available, significantly this does not include white. When white is required this necessitates the base cloth to be a white fabric, and an unprinted area must be left to represent where the white areas of the image are to be located.

Traditionally, in screen-printing, the more colors that are required, the more screens and thus the greater the cost. This is not the case for a digital print in which one color costs the same to output as a print comprising millions of colors. There may be a temptation, therefore, to take advantage of the range of colors available for the same, fixed price, even if the original intention of the designer was to create an image with a more limited color palette (Carden 2013b). The result of this is that limitless color choices can alter a practitioner's way of working by rethinking the creative decisions they make.

With digital textile printing, images can be created in a number of ways, and because highly complex visuals can now be printed that were once impossible to realize, the boundaries of what is now feasible have greatly expanded. However, this brings with it a new set of concerns for designers due to the reduction of constraints that formerly helped them define and shape the aesthetic of printed textile outputs. For screen-printing, designers have to consider the number of colors and complexity of images before translating the individual separations onto silk screens for printing. None of these are barriers anymore and the lack of these constraints on the designer risks a loss of focus and clarity for both images and meaning (Carden 2013b).

Images can be created for the sole purpose of digitally printing on textiles, or they can be selected and taken directly from other digital sources, such as the internet. In each case the translation of the initial visual requires a different form of treatment before it can be printed, and during the translation process the image takes on a different characteristic. Thus the digital to digital translation is less intrusive to the aesthetic content than the analog to digital in which physical to non-physical re-interpretation is required. Going from one medium to another involves decisions that include light sources, tactile qualities, dye options and material considerations, meaning that the final outcome is substantially different from the initial idea (Carden 2013b). This can be a challenge, good or bad, but is something the designer must consider and master. Advanced technology is offering an alternative way of working, an approach to practice that impacts on how designers create, and this is changing how they engage with studio practice when they work with digital printing (Participant 24 in Carden 2013b).

There are specific situations and end purposes for digital textile prints that suit a variety of different forms and complexity of images. For example, the original image might be a handmade visual, a photograph or a digitally created motif, or a mixture of these combined in Photoshop or Illustrator (Participant 21 in Carden 2013b). The designer may also create a library of images that are digital, non-digital or a combination of both processes, and in this way practitioners and clients will be able to benefit from a far wider range of creative options. While experienced designers do sometimes use Photoshop or Illustrator to deliberately create an analog look, usually, if they want a design to appear handmade, then they prefer to do it by hand (ibid.).

If designers did not have a quick access to digital outputs, they would put more thought and effort into the choices they make (Participant 20 in Carden 2013b). This is perhaps one reason why experienced designers are increasingly reflecting on traditional, non-digital processes with a renewed interest and appreciation. They say they enjoy the unpredictability of the outcomes and having to wait with nervous anticipation to see what their final outputs will look like. When something is safe and as expected, this sense of anticipation is lost (Participant 8 in Carden 2013b). A number of designers go further stating that digital is not sympathetic to an artist's instinctive way of working; for example, they almost always start the design process on paper (Participant 9 in Carden 2013b), beginning with a hand-drawn mark (Participant 21 in Carden 2013b); whether it is a pen or pencil, the initial sketch is still made away from the computer (Participant 12 in Carden 2013b).

The physical nature of the non-digital outcome is always evident in the output, not only in relation to the layering of dye on the base cloth, but the associated light qualities, such as opacity, transparency and translucency (Participant 19 in Carden 2013b). The processes used to manipulate materials for non-digital printing impart their own distinctive light-based characteristics on the outcomes, but in digital printing unexpected physical or chemical situations like these do not happen.

When creating with analog techniques and processes a designer can control almost every aspect of their work, but by using digital this makes the practitioner rely on multiple decisions, skills and artistic judgments of others (Participant 24 in Carden 2013b). Loss of control of key components of the design process leads practitioners to think differently about their work, and they must decide which aspects to focus upon, and what to disregard, particularly hard when they have little or no influence over it. For example, for most textile designers exploring through making by hand is an essential part of the creative process; they can see where that process is taking them. They may not have a definite end view in mind, or even what the final outcome is going to be, but the engagement with the materials is fundamental to their creativity and digital can eliminate this stage in the anticipation of easier, quicker and cheaper ways of working.

There is a perception among some designers that they are being judged by the quality of the software rather than on the merit of their work (Participant 23 in Carden 2013b). Software is a creative tool and, like any other tool, designers respond to it and use it in personal ways. This interaction can be perceived as being linked by association to their possession of, or access to, a certain level or standard of software.

Distorting and Manipulating Properties

These opportunities address the concern expressed by many designers that digital textile printing can be overly precise and clinical. Practitioners feel that advanced technology is making them lose touch with many of the material-led senses that they would normally associate with their design practice. By considering unusual ways of approaching the studio experience it is possible to achieve unfamiliar outcomes and promote a greater level of risk in the creative process. The research and development already undertaken in providing the software and hardware for digital textile printing has involved the previously applied skill, dedication and judgment of many experts from numerous associated fields, such as programers, engineers and designers, so any new technology will be adding cumulatively to the work of others.

A single idea can generate limitless variations, and when digital provides unconstrained alternatives it can be excessive in that it alters the creative process bearing in mind it is the technology that produces the variations, not the designer (Carden 2013b). When the creation of this range of images is generated through the technology, and has not originated directly from the designer and their initial idea, these images are actually determined by the nature of the technological provision supplied by the software developers.

The digital technology and its usability has been developed and tested through many stages involving multiple collaborators, so the final versions of advanced technology that come on the market and are available to designers to use represent a group outcome in terms of their components and are thus not tailored to any one particular way of working (Participant 22 in Carden 2013b). These provide a general way of working that accommodates the widest possible range of tools and options for designers but they are not necessarily the best creative solutions for any one practitioner. By attempting to build useful commercial opportunities the functionality of the result is a convergence of creativity and utility, but the reasons for developing the technology are ultimately commercial, not primarily creative. Designers are not usually developers so the software is a response to a variety of needs and requirements but does not fulfil the exact creative demands of any one practitioner.

Digital allows designers to bring a different narrative content into their work (Participant 4 in Carden 2013b). It also means that practitioners can be more relaxed about making mistakes, and, if they are not worried then they react differently to the creative process, designing with less concern for wasted time, materials or the inability to correct things (Participant 1 in Carden 2013b; Participant 12 in Carden 2013b). At the same time though, it is often a heightened anxiety that pushes designers to take chances for an element of risk induces unusual and innovative solutions (Carden 2013b). However, trying out ideas without having to worry about time restrictions or the expense of making something in the real world can, under the right circumstances, also be liberating, such as the testing of prototypes (Participant 1 in Carden 2013b). When we can go back and re-do something, do it quickly, easily and cheaply, it does mean that we do not have to make instant decisions about the quality of something, select only the best pieces or choose what to keep. By putting this is off and knowing they can keep so much that is digital, designers are increasingly avoiding having to make these unpleasant, often difficult decisions about what is of a good or adequate standard to hold onto, and being able to identify or appreciate what should go; being able to make that kind of instant decision is part of a skill that needs to be learnt, and this is in danger of being lost or at best diluted when so much work can be kept or redone later.

PRODUCTION OF DIGITAL TEXTILE PRINTING

Large format printers, such as inkjet printers, use reactive dyes that do not readily penetrate the base cloth. This characteristic of digital textile printing varies depending on the fabric that is chosen. However, if we look at a sample of digitally printed cloth (see diagram 3) we can see that the image routinely appears to skim on the surface and does not seem to fully bond with the base fabric. Once printed, the digital print still requires to be fixed by steaming, and knowing the print can be distorted or removed at this point gives the impression that an image is somewhat transitory and distinct. The idea of a print that sits or floats across a surface can give it a temporary, disconnected aesthetic, and it is the appearance of gliding that is voiced by students when asked to comment on the pros and cons of digital textile printing (Carden 2013b).

A digital print is produced under predetermined controls, such as how the dye is generated, manipulated and then deposited, so it is clear that these are dissimilar to the methods used by designers in an analog printing situation. A digital image can be broken up or unpacked into steps in a particular order and this helps to define the relationships between variables in the process.

Digital printing technology has enabled the workmanship of certainty element to overtake the workmanship of risk component in the printing process (Carden 2013b), and the physical, tactile and sensory risks that used to be implemented within the creative process have, as a consequence, been significantly altered and reduced. What designers working with digital textile printing now find is that the newer processes available to them have to be reconsidered and reworked while many of their preferred visual idiosyncrasies are no longer possible (Participant 8 in Carden 2013b).

As a designer works with, and between, digital and non-digital processes their instinctive creative development in one medium is conducted differently when it is transferred to the other (Carden 2013b). What is undertaken naturally with materials in a studio setting needs to be re-evaluated when attempted in a digital circumstance; this is because the level of physical presence is no longer the same (Participant 19 in Carden 2013b).

Designers have inbuilt creative knowledge, something that is developed over many years, and it tells them that one particular way of achieving something is best for them. This subconscious feeling about how to do things means that their individual, personally developed experience is key to knowing and understanding what works for them (Participant 24 in Carden 2013b). Challenges, combined with skill, then allow these practitioners to try out new ideas within a context of risk, trial and error, confidently knowing that they will learn something useful from whatever outcomes transpire (Participant 21 in Carden 2013b). Using code to design work is another option, and this involves being able to create more complex images while also meaning there is even less wastage of materials and chemicals at each stage of the design process (ibid.). Digital printing, however, can be limiting, for example, there is no option for flocking or foiling. The single setting of, for example, a large format inkjet printer causes the color quality and dye uptake to be problematic on heavier fabrics, and can limit the appeal of the printing process for practitioners (Carden 2013b). Designers often start with analog, use a bit of digital, and then end with digital as a way of getting around this issue (Participant 21 in Carden 2013b).

By providing unlimited possibilities digital is giving designers more choice and freedom but is doing little to control or direct the use of these new possibilities (Participant 4 in Carden 2013b); while analog is limited to materials, digital does not share these constraints. Digital does, however, allow effects that are remarkably detailed and these can exceed expectations in terms of intricacy, complexity of images and multiple layering of narratives (ibid.; Participant 19 in Carden 2013b); and, with advances in technology, these opportunities are accessible to increasing numbers of talented people (Participant 20 in Carden 2013b).

While expectations of digital are different to those of analog, one of the main criticisms is that digital is too clean and feels less intuitive (Participant 9 in Carden 2013b). Designers are deliberately avoiding a generic digital look and are even keen to stay away from any digital tool that mimics a handmade mark or tries to create a fake handcrafted appearance (Carden 2013b). Expert practitioners perceive filters that try to emulate a non-digital process as fake, and even describe them as a way of cheating (Participant 4 in Carden 2013b). Experienced designers who may be embracing digital opportunities are still maintaining a protective ethos towards the older techniques and technologies that they spent years mastering and know intuitively how to use with a high level of skill (Participant 21 in Carden 2013b). Those who possess significant experience have been shown to easily identify superficial mimicking processes (Participant 23 in Carden 2013b). Because these tools are not as natural to use as the ones that took years to master, there is an inbuilt reluctance to use them in their place (Carden 2013b). No matter how elaborate or complex a digital replicating process becomes, a digital mimicking tool cannot fully recreate the function of a human hand, nor its inherent creativity (Participant 22 in Carden 2013b). The act of drawing, for example, requires a three-way looking between the monitor and the page and is less intuitive than a pen or pencil worked on a non-digital surface, and when an action is less instinctive the outcome can never be the same. Regardless of the discipline, designers who use digital methods for printing are unanimous in their dislike of the type of digital processes that try to replicate analog techniques; but, while they may not like these mimicking tools, most recognize the need to use them at some point in order to compete with other practitioners in their particular fields (Participant 12 in Carden 2013b).

Textile designers are increasingly aware of the need to hand over the final printing to someone else, and that person may not be located anywhere near them. Distancing the designer from the means of the printing process not only separates them physically, in terms of contact and procedural knowledge, but also means that they do not personally need to witness the act of printing (Participant 14 in Carden 2013b). This accentuates the digital and physical divide that exists between the designer, the digital image, and the final digitally printed textile. For example, the surface upon which a preliminary drawing may be created is not the same as that of either the end product, i.e. a digitally printed textile, or the means of producing it; this is created in one medium for output in another medium, with the translation being the only direct link between them (Participant 18 in Carden 2013b).

As happy accidents are not possible in digital designs, practitioners need to know before they start what it is that they are trying to get out of the process at the end; they need to understand what the original idea comprises, and what will be expected of the final print, including the dye and materials (Carden 2013b). By the time an idea is translated, all the main decisions must have been made and the designer is no longer in control of the output process; after that point, the practitioners need to trust the pre-set instructions that are to be carried out by the technology (Participant 24 in Carden 2013b).

Analog tends to allow designers flexibility to work more deeply into the process of making (Participant 8 in Carden 2013b). This is not always possible when time and cost are major factors, but practitioners feel that digital restricts their exploration of thoughts and this leaves them feeling that ideas should ideally be worked out first with pencil and paper, rather than jump too quickly onto a computer (Participant 6 in Carden 2013b).

Digital effects are the result of many other people's decisions and experiences, and represent multiple choices made by many nameless individuals over an extended period of time (Participant 15 in Carden 2013b). Many designers feel that digital is still not as efficient as older technologies at replicating processes that were originally and specifically designed for those older techniques (Participant 11 in Carden 2013b). This is because it is not just the output from the original technology that is being copied, but also the relationships between those physical components and processes, and these are far more complicated and complex to replicate (Participant 19 in Carden 2013b). Also, working in analog involves a tactile visuality that is unobtainable in digital, so is a different kind of creativity, and what attracted and stimulated a practitioner in the original context of a discipline has now changed (ibid.).

SPATIAL AWARENESS

The space in which a designer needs to be located in order to create impacts not only on what they produce, but how they feel about the making process. For example, an analog dark room is a solitary, thoughtful place that instils a distinctive kind of approach, whereas a digital darkroom can be anywhere, with no clear physical restrictions, so this particular way of working is no longer applicable (ibid.). For practitioners who have developed an affinity with discipline-specific creative spaces, the loss of these distinctive locations can be unsettling and disorientating and this consequently affects the way they work.

Practitioners who work with analog appreciate the different physical qualities that various materials display towards the designer as well as between the materials themselves (Participant 5 in Carden 2013b). Every material or technique imparts its own identity on the creative process, and the designer's personal way of working blends with this. Digital, however, delivers a one size fits all in that only one dye option is available, only one pressure of dye application is possible, and only one pass of dye is required. The dialogue formed between the designer and the materials in this situation is thus minimal, with the type of dye, quantity of dye, application of dye all significantly different (Participant 16 in Carden 2013b).

Due to increasing availability of digital technology, the amount of analog processing and materials is disappearing, and this is changing what can and cannot be used to create (Participant 8 in Carden 2013b). Designers are increasingly concerned as cheaper and safer options are taking their place, and digital is dictating and defining much of what is now available to them (Participant 20 in Carden 2013b). Digital is emulating analog, although it has not yet fully achieved this goal (Participant 10 in Carden 2013b); instead we are seeing digital becoming truer to itself, more honest and authentic, rather than a poor imitation of an older analog system (Participant 20 in Carden 2013b). Just as traditional effects were designed for older technologies, digital effects are increasingly being recognized as outcomes associated with advanced technology (Participant 11 in Carden 2013b).

While older materials and equipment are being less widely used, they are still valued for what they can do within a niche setting, so what is evolving is a collective of practitioners who strive to keep older equipment maintained and functioning, but within a different context (Participant 20 in Carden 2013b; Participant 8 in Carden 2013b). Such enthusiasm, for example, with woodblocks, metal type or film, results in a deeper understanding of the traditional materials, older processes and techniques because these are being appreciated in a way that was often taken for granted previously, when the materials and processes were commonplace (Carden 2013b).

Digital can achieve many outcomes that were too complex and time-consuming in an analog situation, with ideas now being realized that were not considered possible until recently (Participant 19 in Carden 2013b). However, the support network that enables older technology and materials to work is changing and is thus affecting a practitioner's ability to choose to work with these materials; they may still want to use them, but the cost and availability are restricting them (Participant 20 in Carden 2013b). Experience of working with, or experiencing, a material or process is needed if it is to keep its values and integrity alive; without this, people will not know its true value, or how it can be used most effectively (Participant 19 in Carden 2013b). Analog is more costly than digital and often precious, but these factors alter how people work with it and this is an important issue for their creativity (Participant 20 in Carden 2013b). There is increasing pressure on designers, from all directions, to make things fast with quick decisions. Such expectations require speed, rather than slow contemplation, although the upsurge in slow design is challenging this stance in certain areas where there is appropriate contextualization and marketing to support it (Carden 2013b). As more technology and machines release time for people, they have more time to champion the benefits of additional reflective time in slow processes (Participant 20 in Carden 2013b). Expertise that took many years to develop is still needed by designers because practitioners who spent their whole lives learning a craft are the ones best placed to uphold the values of a discipline that they deeply and fully understand. Although digital challenges this need for in-depth knowledge, what is often produced is the result of non-embedded knowledge, superficial talent that leads to problems in ongoing development because in this situation it is not possible to build on increasing levels of skill (ibid.).

PROPERTIES OF DIGITAL TEXTILE PRINTING

The meaning of digital textile printing can be divided into three categories: the generic properties of digital printing, the experiential knowledge associated with digital printing and the transferrable knowledge linked to digital printing.

Generic Properties

Traditional hand printing enables a designer to maintain creative control during the entire printing process, while digital printing requires the practitioner to pass over control of this key stage as soon as the digital file is ready to go to the printer.

Techniques regularly dictate the options available to designers and, in turn, these impact on the visual of the final outcomes (Participant 24 in Carden 2013b). Influences include software and hardware provision as well as cultural associations absorbed and formed from the world around us (Carden 2013b). Even our subconscious tells us if a textile is digitally printed or not, and this is reinforced by several generic properties of digital printing, for example, when digital dictates that a white fabric is used if any white color is to be included in the final print.

What is a digital print? If the image is, or is translated into, a file then it is digital; if the process used to print the base cloth is a digital printer, then it is digital; so, if the image goes through a coding stage and, or, a digital printer at all, if it has a digital component within its composition then it is a digital print.

The digital print possesses a smoothness that looks devoid of hand input (Participant 20 in Carden 2013b). This precision can make digital look mechanical, a factor that feels uncomfortable, especially for a product intended to be worn against the human body; there is also something uncomfortable about bonding or combining a human element with a machine in such an intimate manner (Carden 2013b). A non-human relationship exists between the application of dye and the fabric when there is no evidence of handcrafting in the visual aesthetic of the output (ibid.).

When a designer creates a handmade print there is a physical and emotional component inherent in the output. This is caused by the practitioners' background that they have brought to the process that day, and this includes responses to the way they might be feeling, pressures they might be under, stress or emotional turmoil, issues that impact on how the output of creativity will look; in additional, the material, the fiber or its history, the type of fabric and the atmospheric conditions under which it is stored or printed will all influence and be evident in the final artefact (ibid.).

A lack of constraints makes practitioners work in alternative ways, and digital affects the way designers who learnt their skills in an analog situation, now work with advanced technology (Participant 4 in Carden 2013b). Less and less constraints cause creative boundaries and definitions to be changed because there are so many new options available to them (Participant 5 in Carden 2013b). On the one hand there are limitless creative avenues open to them, but in reality there is a consistent even look to the quality of the outputs they produce (Participant 8 in Carden 2013b). The range of images is vast and also how they can be varied; this may be far more quickly, and easily, but more people can access the same advanced technology, whether they are skilled or not, so it is harder to separate the good from the bad when there is simply so much stuff out there (Participant 20 in Carden 2013b).

Digital is not the same as analog so it should not be viewed as such (Participant 14 in Carden 2013b). However, the problem still remains that visually, digital can make things look more skilled than they possibly are, and the polish that it provides can be deceptive (Participant 1 in Carden 2013b). Designers are only too aware that executing work in a digital context can mask a gap in skills in the analog world, so it is the meaning that underlies the design process that is key to further successes (Participant 2 in Carden 2013b). Increasingly designers are approaching tasks knowing what they want to achieve (Participant 12 in Carden 2013b); and, anticipating in advance what you want to get out means that traditional methods and skills are increasingly important because no working out can take place once a practitioner gets started (Participant 11 in Carden 2013b).

With digital process what we get can be more of a record than any interpretation because the act of printing is not an integral part of the making process any more (Participant 19 in Carden 2013b). The tactile response associated with creating and making is gone due to digital removing touch from the imagination and interpretive situations, and, because digital is very right or wrong, there is less margin for experimentation and this limits the range of acceptable outcomes (ibid.), Clean is not always aesthetically desirable, but this is something digital does better than analog; because it does precisely what it has been predetermined to do (Participant 20 in Carden 2013b).

Experiential Knowledge

Designers use materials and techniques in ways that are, for them, as honest and personal as possible. The availability of packages that mimic skills that would ordinarily take a considerable length of time to master, challenges the existing process for gaining experiential knowledge, and this is problematic for practitioners (Carden 2013b).

Personal practice involves a culmination of skills and associations, previous knowledge, cultural experiences, memories and so on, and how we respond to technology and digital outputs plays a large part in our decision-making, and this changes and naturally evolves over time. As the technology used in digital textile printing develops, both in terms of what it can be made to achieve and in how practitioners design with it, the increasing reduction of human involvement alters the nature of the final aesthetic (Participant 18 in Carden 2013b). However, the increased availability of advanced technology brings with it many of the benefits of workmanship of certainty but this can also be at the loss of many of the benefits of the workmanship of risk (Carden 2013b).

By knowing the nature of what a designer wants to achieve, especially by manipulating and playing in a physical capacity with materials, it can be possible for practitioners to reflect and better understand what the relationship is that they have with the various materials they use. Thinking about what to do only gets a designer so far; it is necessary go on to experience the physicality of materials, to play with and try out various juxtapositions of color and texture (ibid.); none of these are possible on a computer screen (Participant 24 in Carden 2013b). Digital changes the relationship designers have with materials when it offers preplanned options that the advanced technology has been programed to provide. For example, controlled digital randomness can only operate under predetermined parameters because a machine cannot think outside the boundaries it has been set (Participant 22 in Carden 2013b).

Practitioners take longer to become familiar with older technologies and materials, and this lack of comfort is compounded when they hear stories from other designers about expensive accidents that digital can now help to avoid (Participant 1 in Carden 2013b). As non-digital requires a greater level of care, it is harder for designers, especially younger or inexperienced ones, to aim for the magic of analog when they can rely on the safety of digital (ibid.). However, experts know when the computer is appropriate, and what areas are best handled in an analog situation (Participant 21 in Carden 2013b); for example, in film production, analog archiving is preferred over digital for materials-based creations (Participant 20 in Carden 2013b).

Some handcraft textile designers are concerned that even though there are infinite possibilities and millions of colors available to them, they are finding that this often means less scope because the physical and material opportunities they like to use are more limited in digital than with analog processes (Participant 24 in Carden 2013b). For example, the texture of hand-woven fabric is often unsuitable for digital printing and recycled cloth may have been subject to chemical treatments that hinder the application of reactive dye. Analog is also better understood and is used according to more established rules and guidelines than digital at present; although this situation may change as digital becomes more embedded in the designer's toolbox (Participant 23 in Carden 2013b).

Transferrable Knowledge

Utilizing advanced technology alongside non-digital processes, especially those shared with other disciplines, allows designers to build theories and meaning from areas of commonality (Carden 2013b).

To help designers work efficiently, it is necessary for ways of interpreting intentions behind the creation of digital images for digital print to be more transparent. The various senses mean the outcome will look different; emotionally it will be different; the physical nature of an outcome becomes far removed from the medium of the original idea, and so the senses could be the variables that inform the translation of the idea or image across different mediums and on through the analog to digital design process (ibid.). For example, not requiring a handcrafted element means that digital printing has lost its capacity to evidence the hand that made it (ibid.).

While advanced technology is constantly evolving, so are the materials and fabrics available to practitioners, and the prints from the past no longer look the way they once did (Participant 23 in Carden 2013b). For example, designers react differently if a base cloth is synthetic, or natural, recycled and ethically sourced and produced. After a period of time our relationship with textiles changes; at first an image is contemporary and the textile is freshly manufactured, but gradually the image is dated, possibly faded and worn, but then memories kick in and the textile acquires a further, personal and unique meaning (Carden 2013b). At this stage, textiles can become empowered with associations and references that give them a value beyond what was initially created. Designers are aware that the output they produce might look good, but if this is not backed up with sufficient knowledge about how this aesthetic was achieved, then the information needed for the practitioner to recreate this success will not be immediately available to them in future (Participant 22 in Carden 2013b). Everybody can produce lovely things with digital tools, but then what? They need to know what makes it good, what sets it apart as creative and better than something else that has also been produced with similar advanced technology and is also out there (ibid.). Designers acknowledge that using digital can be easier, meaning that the requirement to master a process is less necessary, but while learning a skill takes time it also gives the practitioners a chance to try, to err, to learn from their mistakes and to play, and this is lost when that requirement is taken away. What this means to designers is that they create differently (Participant 15 in Carden 2013b). It also means that they can make more of something, but again it takes experience to appreciate the quality of something and selection is itself a skilled activity that takes time to develop and master (Participant 22 in Carden 2013b).

A further consideration is that a smaller amount of time is now used on developing skills and less time is spent on decision-making, meaning that less thought has been put into the choices practitioners make (Participant 20 in Carden 2013b). For people who are not an expert it can be a case of making something that previously they might not have had the talent, or access to the necessary equipment, to do in an analog circumstance but now, however, almost everyone can create something in digital. This does not mean that every person is an expert, so it is even more important that those who have mastered a domain can still judge if a piece of work is good or not (ibid.). Another criticism of digital printing is that it can be perceived as cold, not engaging, less human and impersonal, so having previously gained skill is again important in ensuring that such issues are challenged and addressed by making the best use of advanced technology and the available opportunities (Carden 2013b). Designers who have grown up with analog know subconsciously what to expect from a discipline (Participant 20 in Carden 2013b). Initially digital can seem a quicker, cheaper way of producing things, but in reality it creates differently, and increasingly practitioners are spending time trying to get digital to look more like analog (Participant 10 in Carden 2013b). When digital was not an option, people spent time learning alongside older masters and acquired a great deal of knowledge by being close to someone while they worked, picking up nuances, a feel of how things worked and what created the best results, procedural knowledge, factors that can only be picked up by being close to someone who has spent years with a technique and materials. Digital involves a great deal of knowledge from a wide range of experts but much of it is contained within the hardware and software of the advanced technology, and unfortunately those people are not present when the machine is in operations and are therefore unable to pass on all their knowledge, so in this respect the advanced technology cannot pass on this type of knowledge to inexperienced designers (Participant 20 in Carden 2013b).

Older materials, techniques and processes, plus an audience's expectations, are all relevant to contemporary practices because if something is made to look old it usually looks fake (Participant 23 in Carden 2013b). This is mainly because the materials, techniques and processes collectively are not right for the look of the time they are intended to portray, and an expert audience can usually tell. This highlights a further problem: as with an ancient language, if designers lose the skills, techniques and regular use of a traditional analog system they will not be able to skilfully return to it in the future, because once the knowledge and background support for it have gone it may be impossible to find them again (ibid.).

PRACTICE OF DIGITAL PRINTING

When attempting to create textiles that do not conform to a uniform or predictable digitally printed look, it is first necessary to ascertain what is involved in creating a digital textile print.

The practice of digital textile printing comprises two themes: physical characteristics and the impact of processes.

Physical Characteristics

Designers state they feel excitement when they work directly with yarns, fabrics, colors, different tactile qualities, layering and intertwining them to build structure, color combinations and texture (Participant 20 in Carden 2013b; Participant 21 in Carden 2013b). These characteristics are fundamental to the way textile designers work, and traditionally these have been key to how practitioners operate in their studios. However, the natural physical conflicts of multi-sensory material thinking are no longer available when designing for digital textile printing (Carden 2013b).

Most designers instinctively draw by hand to start the creative process. Because paper has texture, and thus a natural resistance, when practitioners move a pen or pencil over the surface of the paper they are responding and interacting with the tactile qualities of the paper and the physical traces of the coloring material. As the work progresses, visual and physical senses pick out detail; they can feel the variations in the thickness of the surface and the corresponding impact of the mark making tools as they are doing it (Participant 24 in Carden 2013b). Even the process of setting up the surface, the activity of preparing the material that will be used for creating something, is part of the mindset of creativity because it allows us to mentally and physically prepare ourselves and to start thinking about how and what we are going to make. This physical act takes time: it is a ritual that helps us to anticipate how we are going to create and is lost in digital circumstances (Participant 22 in Carden 2013b). Other than switching on the computer, there is little comparable preparation work required to start creating in digital (Carden 2013b).

For designers, the more precious the raw materials, the more fraught with risk the making process; this can be due to cost of the materials, or time constraints, or an inability to go back and re-do something, but whatever the reason, designers agree that the elements that cause them to take chances actually help them make extraordinary decisions (Carden 2013b).

Sketchbooks are fundamental to textile designers and these enable practitioners to freely, privately and unrestrictedly put down partially formulated thoughts, work up rough ideas, experiment with initial drawings, or write a few words on paper (Carden 2013b; Participant 21 in Carden 2013b). A benefit of sketchbooks is that ideas can be quickly recorded in a very personal style, be readily revisited, can comprise different media, be rough, be textural, may demonstrate handwritten notes and cross-media associations, but primarily they reflect ideas and thoughts in such a way that the designer still has total control over their creativity (Participant 24 in Carden

2013b). After the sketchbook, everything can be brought together in Photoshop or Illustrator (Participant 21 in Carden 2013b).

Impact of Processes

What seems to stand out when working with advanced technology for textile printing is a machine look, smoothness and uniformity; this is what most observers associated with digital printing and this is not usually a desirable visual aesthetic.

When working in digital other people have made the decisions regarding the composition of the dye, the type of dye, the size of the print heads, the exact pressure used to force the dye from the print heads onto the fabric, the height of the print heads from the base cloth, the size of the inkjet printer, the width of the printer and thus the maximum width that the cloth can be, the type of fabric that can be printed, the thickness of cloth that can be printed, where the digital printer is located, the price of printing, when printing can take place, and so on. All of this means that a designer is not in control of a large number of the processes associated with digital printing (Participant 24 in Carden 2013b). There are many options for practitioners, but also multiple restrictions because of the complexity of the technology and everything that supports it. Instead of the work of many people being combined to make a traditional printed textile, such as a chintz, digital textile printing uses the work of multiple others transferred into technology and a designer has to make choices using the embedded knowledge from these other, non-visible, people (Carden 2013b).

DIGITAL PRINTING PRACTICE IN ACTION

Digital printing practice in action consists of three themes: unexpectedness and variability, pushing possibilities for digital printing and combining digital and non-digital processes.

Unexpectedness and Variability

It is common knowledge that digital printing is not produced by hand so there is little expectation that a handcrafted element will be present. However, while acknowledging that digital prints are machine-made, there are still opportunities for designers to achieve alternative types of outcomes by intervening in the digital process to create multiple narratives and random outcomes (ibid.). Examples include applying pre-printing solution by hand to give a negative batik-type effect or pre-cut shapes can be placed between the print heads and the base cloth to create stencilled interventions in the digital printing process.

Pushing Possibilities for Digital Printing

Using different weights of base cloths, types of fiber and structure can result in a variety of outputs, while alternating the weights of fabric, the fiber content of the fabric, changing the height of the print heads, experimenting with the pre-printing solution in selected, hand-painted areas, plus intertwining handcrafted processes whether individually or in multiple-operations allows the

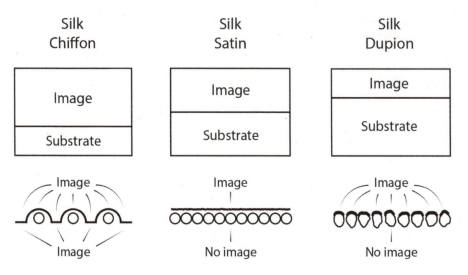

Figure 10 Visibility of digitally printed images on silk. Susan Carden 2012.

designer to mix and match in order to generate a wider array of textile samples. In this way they can take control from within the boundaries of digital textile printing (ibid.).

When looking at six different fabrics digitally printed with a single image and using the same reactive dye, a range of outcomes is possible, and the physicality of light and its consequences seem to be important factors here (ibid.).

Thick cotton: The graining of the weft, and to some extent the warp, causes the image to break up slightly. It is almost as if the structure of the cotton base cloth, though fragmented, is integrated with the image, both visually and texturally. The dye looks to be absorbed, rather than sitting on the surface of the cotton fabric (ibid.).

Thin cotton: The dye is almost not visible on the back, so, unlike a cloth such as silk chiffon, it is not the same on the underside. However, the plain weave of this fabric is light and less densely packed than the thick cotton and almost gives the chiffon's sense of high percentage yarn coverage and means the image is substantially integrated. The image also has a slight graininess that allows the base cloth's texture and presence to be visible. This integrates the image of the print and makes it appear to be somewhat saturated into the fibers of the cloth, rather than skimming on the surface (ibid.).

Silk chiffon: The image does not look to be floating over the surface, nor to be adhered as a separate unit to the top either. The image does not obscure the cloth beneath; on the contrary, it seems to be completely integrated with and into the silk fibers (ibid.).

Silk satin: The surface is relatively even, compact, smooth; as a result the print appears to be solid, less broken up on the individual yarn ends of the base cloth. This means the fabric underneath the image is less evident than the printed visual, and the image seems more separated, individual and distinct; the image is one part, and the base cloth clearly appears to be another. There is not much that contains evidence of both; the print hides the base fabric (ibid.).

Silk dupion: The dye sits across and around the sides of the individual yarn ends, giving a three-dimensional effect and distorting the luminosity and surface of the printed image. This integrates the image with the base fabric more than the silk satin, and thus makes it look less like it is floating on the surface (ibid.).

Viscose: The base cloth is compact and so the digitally printed image is fairly unbroken. However, it is quite dull compared to the silk satin due to the light being deflected to the right and left at different angles to each other, resulting in an overall lackluster appearance to the image. It is also significantly darker, even with the same dye, inkjet printer and source image (ibid.).

Combining Digital and Non-digital Processes

By combining digital with non-digital processes the designer is joining one set of experiences and expectations with another. Making by hand requires the senses of touch, smell, sight and sound, whereas working with digital printing uses only visual. When digital and non-digital are combined this allows the designer to add a greater sensory value to the digital printing process and they can utilize the benefits of both techniques, thereby creating more efficiently for maximum sensory effect (ibid.).

SUMMARY OF CHARACTERISTICS

The digitally printed image seems to skim or float on the surface of the base cloth, resulting in outcomes that are often perceived as fake. When a printed image is distinct from its base fabric, the result is understood to be less authentic.

Designing textiles with advanced technology reduces opportunities for handcrafting and as a consequence outputs often contain a generic digital aesthetic. The more advanced the technology, the less evidence there is of human involvement in the final outcomes.

While digital can make textile designers confident in decision-making because they can easily and cheaply correct any errors, it also removes the element of risk that often provides interesting results. Although digital can enable practitioners to be less worried about making mistakes, it usually requires an outcome to be predetermined at the outset, so there is less scope for experimentation.

The more digital tries to emulate non-digital textile design practices, the lower its perceived value. An expert audience attributes less worth to digital when it is seen to copy analog.

A combination of digital and non-digital textile design techniques results in greater creative freedom and novel outcomes. The joining of digital and analog processes provides synergistic opportunities for practitioners and outputs.

There are changing demands on textile practitioners due to issues with accessing and processing older equipment and materials. Designers are reflecting on older technologies, thereby making advanced technology work more naturally for them.

CONCLUSIONS OF CROSSING DISCIPLINES

Practitioners find that using advanced technology is easier, cheaper and quicker than previous ways of working, for example, it eliminates the need to spend years acquiring specialist skills. This means advanced technology allows them to produce skilful looking results more easily, cheaply and quickly. However, there are also many drawbacks, including a perception that final artefacts have a reduced handcrafted aesthetic, and the outcomes often display an undesirable generic digital look. Artists and designers also feel that the use of digital restricts the way they work in the studio meaning their outputs can often be predetermined, demonstrate unnatural characteristics and are not fully showcasing the qualities of the materials they are using. So while practitioners have confidence that their outcomes of practice are uniformly precise, that these can be easily corrected and reproduced, the technology has eliminated many of the elements of risk that often lead to the most interesting results.

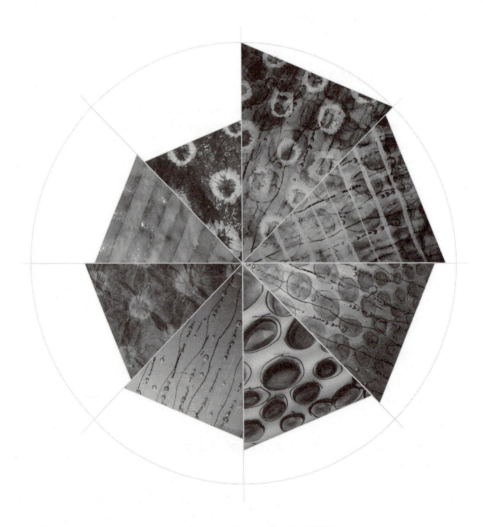

Figure 11 Development of a digital textile print. Susan Carden 2011.

9
CONCLUSIONS

This book has considered how digital textile printing originated, and explores why designers create the way they do with this relatively recent printing process. By exploring how each stage evolved and developed, the constraints and conflicts that informed or shaped them, including technological, material and visual, this contextual legacy can be used to better understand and explain where the essence of a digital print is located. An appreciation of what the processes embody and where they are situated within a contemporary context will allow artists and designers to create digital textile prints more effectively, encourage them to work with advanced technology in a more natural manner, and promote interactions with older technologies and alternative materials.

To produce a printed image requires pigment that is sourced, ground to a powder, made into a paste, and applied either by hand, by blowing or via a tool. Thus the hand stencils were created by using their own hands as templates and spraying by blowing either across a hand-cupped pigment solution or by blowing the thin paste through bird bones towards the hands placed against the wall. Drawing a hand shape without using a template would have been a far more difficult operation and would have required far more skill and dedication, not to mention judgment and reflection. It would also have necessitated the retention of the idea of a hand and the person would have to start in a selected spot and try to recreate the impression of a hand, something they would have in mind, but through a physical act of working, in real time, bit by bit along the drawn line, anticipate how far to continue the line to create something similar to a three-dimensional form that extends past the wrist and on up the arm. The lengths of the individual fingers and the thickness of the wrist would all have taken much practice and determination to achieve and have involved a leap into the acceptance of the abstract. This would have been a daunting task, and given that many hands were readily available and could be used as multiple stencils, no wonder the image of a hand was depicted in this way for so long.

The availability, location, type and composition of the dyeing material were key to the emergence of this critical mass, while the tools, previous knowledge, skill, drive and emotional conflict, reflection and constraints allowed ideas to be converted into visual forms. Our instinctive desire to understand and interpret the environment around us is one of the earliest indicators of our cultural beginnings, from gathering and arranging shells, to utilizing the human hand to cup, hold and manipulate materials and tools 1.4 million years ago (Marzke 1997; Ward et al. 2013). The fact that our ancestors could use their own hands and surrounding materials to make versions of what they saw can be traced back as a pivotal period in man's development, and helped to differentiate modern humans from Neanderthals and the rest of the animal kingdom.

Once humans could manipulate by hand they began to twist and spin fibers into yarn, they learnt to weave and felt and create fabrics that could be worn, removed and fashioned into garments or shaped for interior spaces. The processes they had developed for cave art, such as spraying, smearing, painting, and stamping, could now be applied to these fabrics. The only two of these methods that are suitable for multiple reproductions are spraying over a stencil or stamping with a pre-formed shape saturated with dye or pigment. Printing by stamping with a previously crafted shape, like a felt cut-out, comprises many individual processes and demonstrates a significant evolutionary step forwards. The image produced by the cave artists was initially a negative hand stencil, not so much a drawing as producing an image by resist spraying. Therefore this technique shows contact with the original hand and, even after tens of thousands of years, still acts as a direct link to the person who made it.

We have looked at woodblock printing, with its abstract images and divisions of labor into three skilled processes: the artist, the carver, and, the printer. All three processes create distinctive characteristics and enable each person involved to develop and showcase their own personalities through their contributions to the work. The materials also influence the final outcomes. The viscosity of the inks, whether wood, metal, roller or screens are used for printing, relief or indented impressions all result in different final prints. Screen-printing with tiny dots of color squeezed through a mesh is similar to the minute 2,304 squares used for the first letterform descriptors, and the pin-prick dots used to create tattoos, and also the pixels on a computer screen all subdivide the unit into far smaller sections that can be used to construct a color separation, whether letterform, tattoo or digital image.

Tools imply patterns and motifs; spraying over a stencil in a cave, using rake patterns for tattooing, color separations for woodblock or metal plate printing, wood and copper engraving techniques, woodblock and copperplate repeats using dots for registration, all give characteristics that reflect the tools and techniques associated with them. The digital designer does not need to be present when an image is being printed and the image can be as complex as the practitioner desires, so the lack of constraints imply the characteristics that define a digital print, for example, unlimited colors except white, images of infinite complexity, and, production decisions that must be predetermined before printing begins.

By identifying connections to the earliest evidence of visual culture, including expressive forming and crafting by hand, it has been possible to link the tools and techniques, material interaction and physiological engagements, skills and the acquisition of knowledge, with the dissemination and social engagements that define culture. These initial circumstances helped build networks and enabled creativity to flourish and expand in far-flung locations, such as South Africa 100,000 years ago (Henshilwood et al. 2011), Indonesia 40,000 years ago (Aubert et al. 2014), America 6,000 years ago (Perez-Seoane 1999; Clottes 2010; Simek et al. 2012) and Europe 4,800 years ago (Pike et al. 2012), and explain how these early skills and experimentation informed and encouraged the act of printing, through a series of different visual processes and cultural influences, to the digital textile printing of the present day.

What digital textile printing does, however, is to allow artists and designers to see beyond the constraints that came with each version of the earlier processes, such as the large areas of flat color needing to be packed with felt for woodblock printing, or the etched shading required to cover similar large areas of flat color, or foliage and vine images to disguise the pattern repeat, or the image of seed heads placed to hide the registration marks. Digital textile printing has no such constraints for repeats, no limits of the types of images that can be created, nor any requirement for color separations; therefore most of the production constraints that influenced and as a result shaped the aesthetic of textile printing have been removed.

While digital textile printing does not possess these restrictions, in a similar manner to the negative hand stencil, it is the very nature of what it does not do that defines the essence of what it is. For example, the smoothness of a multi-colored print with no overlaps, where one dye color nudges up to and slightly overlaps another, is not there. White cannot be digitally printed so the base cloth must be left un-dyed if white is required and this often makes the image appear to sit as a unit surrounded by a white border; because there is no need for individual screens or color separations, more than twelve colors usually indicates a digital print. No restriction on the complexity of the image means that a photographic visual or highly complicated collage normally implies a digital print. No limit on the scale of repeats means that if there is a large image with no repeats, it is usually a digital print. Regular incised patterns or finely detailed cutwork, particularly on synthetic fabric, sealed at the edges with no fraying, also imply digital printing by laser cutting. Uniformity and smoothness of a finished textile print in which the image seems to skim on the surface is regularly described as a digital look. So, invariably, it is possible to tell if a textile print is a digital textile print by identifying what it is not. However, digital textile printing can also enable designers to work quickly, easily and cheaply to create extraordinary outcomes. By incorporating the processes with elements that help to make them distinctive and unique, practitioners who use advanced technology are already changing the way digitally printed textiles are being produced, used, understood and appreciated.

If a digital textile print is not digital, is not a print, and is not defined by the fabric, then how is it a digital textile print? As Dewey (1934) explains, an artefact is a fake when it can be seen in isolation from the creative activity that was used to produce the original. Thus digital textile printing, it would seem, has a number of challenges to face, not least the perception amongst many that its outputs can appear to look fake.

Unlike most previous printing techniques, in digital textile printing the materials and nature of the construction of the base cloth have little bearing on the characteristics of the final printed image. The digital image can come from almost any source, and is not governed by constraints associated with any particular type of printing process, such as color separations or requirements of woodblocks, so has no heritage to guide it. The digital printing process has no human features, it is pre-set, the dye is already stocked, the print heads are fixed in place, and it sprays a predetermined quantity of dye at a set pressure in a matrix formation that has already been programed. Thus the final artefact has, as Dewey would suggest, been produced in isolation from all the various operations that are used to create it. By its very nature it can be viewed as a fake object. But this would be to overlook the difference that exists between something that we know to be artificial or fake (Huang et al. 2011) and a thing that is artistic and has been created not to simulate something else, but as a result of combining together new opportunities that have little history and fewer boundaries.

As Hughes (1999) states, a medium dictates what can, or cannot, be achieved, and this is due to its constraints. Even though the computer eliminates many constraints, it is not able to make everything easier, cheaper and quicker. But as Hughes also points out, it does make some things particularly easy. The issue for designers is not to focus solely on these features but to look for inspiration and new approaches in the areas that are not easy. The case studies in Chapter 5 demonstrated how a number of practitioners are finding inspiration by reflecting on the key techniques of earlier printing skills and processes. Each person uses an aspect, or area, from within the digital textile printing process that they have successfully managed to manipulate or rework according to one of the traditional crafting practices. They have each achieved this in a personal manner for their own creative purposes. Hudson explores the natural and imprecise characteristics of a wool yarn while 3D digital printing with one of the oldest textile constructing processes, felt making; Russell creates personally designed and programed images for textile printing, coding with one of the earliest artforms, drawing; Rigley manipulates images by pulling together sourced material that can only be separated and reconfigured on a computer in a process that pre-dates the alphabet, visual narrative; Goldsworthy uses laser-printing technology to reuse synthetic fabrics that would otherwise have been destined for waste, as she has developed a method for creating digital laser-cut patterns on reclaimed fabrics that reinvents both recycling plus the ancient embroidery technique of cut work; Treadaway looks at the implications of creating an alternative visual language in a non-physical space, with non-global color fidelity, taking artistic practice and translating as images through 3D printing to achieve a digital version of an early tool-based skill, carving; CAT uses digital textile printing to make on-demand prints of archived material that is no longer in production and thus normally unavailable, creating a unique textile piece for someone; Britt and Bremner take archival inspiration and re-interpret through the digital process to output, generating image creation with digital tools; and Carden has revisited the traditional techniques of fabric resist dyeing, including batik, tie-dye and ikat, and has manipulated the properties of the digital textile printing pre-coating solution and printing process to create alternative versions of resist dyeing. These practitioners have therefore reinvented the traditional textile practices of felting, drawing, visual narrative, cut work, carving, spraying, image creation, and resist dyeing for use with advanced technology.

While digital textile printing eliminates the need to have skills associated with the creation, preparation and production of the textiles, having those skills enables designers and practitioners to be able to use digital textile printing in an informed manner. It allows them to knowingly make decisions about what they do and how they do it, and it also gives them the opportunity to make advanced technology work for them in accordance with their own individual creativity. They know from personal experience as textile designers what features and characteristics of their design practice they wish to explore and what aspects of the digital provision they can and need to manipulate for their own work.

Without deliberately and intentionally going into the process and investigating what can be done differently, and more creatively, the digital prints could comprise almost any selected existing image, output on any number of colors on any of a vast range of existing base cloths. Being a creative textile designer is not, and should not be, just about selecting an image (Russell 2009) and choosing a fabric.

Significantly, each of the practitioners would not have been able to produce the outcomes detailed in Chapter 5 without advanced technology. So not only are they able to play with technology through their own personal creativity and skill, but they are ending up with outcomes that could not have been produced previously in any other way. But even more important is the fact that had they not been skilled and knowledgeable about the traditional processes and techniques developed over hundreds of years by previous textile designers and producers, the ones they were able to manipulate and introduce into the digital textile printing situation, then they would not have had the necessary mastery, procedural or intuitive knowledge to undertake such work in the first place.

In designing with advanced technology, digital textile designers are beginning to lay the foundations for what this exciting new medium represents. However, doing something when you can do almost anything is difficult. Also, being able to design and produce while remaining detached from the means of production is possibly the biggest challenge for practitioners working with digital textile printing, but this may also represent its greatest asset.

By observing digital textile printing from the present day back to its prehistoric origins, in terms of both process and practice, a connection can be traced between the act of handcrafting with a previously sourced and manufactured coloring material and the use of a previously considered and organized template, in order to create an image. This is a three-way action requiring a person with an image in mind, their hand constituting a stencil and that same person's directed blowing power producing the final outcome.

With digital textile printing, this three-way action has been fragmented so that first the designer, second the means of printing, and third the materials used, no longer act as one unit. Authenticity is also more difficult to ascertain once the division of labor takes place. In digital printing authenticity lies in the space that exists between the moment the digital image is sent from a computer to the printer, but stops before the dye that is ejected from the print heads reaches the base cloth. This means that the boundaries of authenticity for digital printing have evolved as labor has distributed, and are no longer the same as those of the three-way process that was used to produce cave art.

Digital printing does not require the designer to be present during printing. The designer need never come into contact with any of the materials used to produce the image, neither before nor after printing. This means all three actions are no longer bound by physical human contact thus changing the nature of the communication connecting the image, the material and the process and the message that the digital textile print transmits.

Digital images can easily, quickly and cheaply be copied, especially for fast-fashion. In order to regain control of authenticity and the creative process an artefact needs to recreate a link to a person, or group of individuals and increasingly digital textile printing is about evidencing the human input through handcrafting. This is imbuing artefacts with skill and uniqueness, characteristics that are reflected and valued across the spectrum of disciplines that use digital printing, from letterpress to textile design. As a result, contemporary designers are instinctively reflecting on the skill of handcrafting to redefine the message they are communicating through the medium of digital printing.

GLOSSARY

Defining the key terms used in the field of digital textile printing helps to explain the individual techniques and processes involved, as well as the context for their application.

Digital textile printing
The individual component terms of digital textile printing appear to imply processes relating to, or performed by, human hands. However, digital textile printing as a combined term, can generally be understood to describe contemporary printing methods that process digital files rather than, for example, woodblocks, copper plates or silk screens. The term also refers to the act of transferring a digital image directly onto a base cloth, so is partly in the material world and partly in the virtual one. In broader terms it includes a variety of methods for using electronic files to produce an image composed of dye or ink. Also, due to the image being drawn anew for each application, rather than requiring a secondary contact material, such as a woodblock to be shaped, carved or prepared beforehand to carry the coloring agent to the cloth, no test samples are needed. However, digital textile printing does not produce an impression or stain, nor does it come into contact with a base cloth, so technically it is not a conventional form of printing (Campbell 2008). So, for practitioners who are skilled in using their hands to press, apply color and construct, digital textile printing can be something of an anomaly, as it does none of these things.

In broader terms it includes a variety of methods for using electronic files to produce an image composed of dye or ink as dots. Also, due to the image being drawn or applied anew each time, rather than requiring a secondary material, such as a woodblock, being carved, shaped or prepared to carry the coloring agent to the substrate, no test passes are needed either, and therefore encompasses process for producing digital prints using CMYK combinations for each individual dot of color, allowing a digital image to be interpreted in a single application or pass, of millions of colors (Braddock Clarke and O'Mahony 2007; Harrod 2007; Ujiie 2006).

Other key terms used in this book include:

Adjective dye
Adjective dye is a category of dye that needs to be used in conjunction with a mordant, such as alum, tin, chrome or iron, in order to fasten the color to the fibers at a molecular level (Dixon 1979).

Advanced technology
Advanced technology can be defined as the practical application of knowledge in relation to digital, rather than non-digital, systems (Bunnell 2004; Flusser 1999; Hopper 2010; Manovich 2001; Treadaway 2007).

Analog or non-digital
Analog or non-digital is a term for that which is produced, or reproduced, without the direct involvement of advanced technology (Althusser 2008; Eco 1989).

Batik
Batik is a resist dyeing technique that is applied to a fabric once it has been constructed, rather than for example ikat, which is a yarn resist dyeing process. Wax is traditionally used in batik and the melted wax is poured in a pattern onto the base cloth. Once cooled, it is then dyed and the wax is removed by heating, usually with an iron, between two layers of absorbent paper (Robinson 1969; Kerlogue 2014).

Chintz
Chintz is another name for the popular colorful Indian dyed and printed calico Indienne. It requires multiple processes, including pricking out of the fabric, woodblock printing, dyeing, resist dyeing, and painting. Numerous skilled craftspeople were involved in creating chintz textiles and a distributed network of workers was needed to undertake the various tasks in different locations or houses. The intricate decorative patterns were mainly used for upholstery and curtains (Crill 2008; Riello 2010).

Communication
Communication is the act of conveying meaning involving at least one of the senses (Leach 1976; Frayling 2011; Williams 2012).

Constraint
A constraint is a restriction that is placed on someone or something, but it can also help to shape or govern a discipline (Gerber 2004: Casey 2012).

Copperplate
Copperplate is a printing process that was developed in the fifteenth century. Although more expensive than woodblock, copper takes longer to carve but lasts longer and as is less prone to wear. Unlike woodblock cutting in which the artist was traditionally unlikely to carve the image, for copperplate the engraver was often a goldsmith or silversmith and normally engraved and even printed his own artworks. For textiles, an image is cut into the surface of a sheet of copper, similar to an etching process, ink is spread across the surface and wiped, leaving dye in the cut lines. The design is then transferred to the surface of the fabric (Chamberlain 1978; Russell 2011).

Cotton
Cotton is a plant that has been grown successfully for textile production since ancient times in numerous locations around the world, including India, China, Indonesia, Central Asia, Turkey and Egypt. The cotton plantations of North Carolina and New England were instrumental in defining

the culture and economy of the American states, and cotton manufacturing helped to promote the industrialization of the textile industry. The popularity of cotton was crucial to the development of trade between East and West and also the spread of dyeing techniques, pattern designs, fashion, cultural commodities and ideas (Lemire 2011; Riello 2013).

Craft
Craft can refer to the application of skilled work and is generally associated with particular physical materials (Lees-Maffei and Sandino 2004; Pirsig 1974; Veiteberg 2005).

Craftsmanship
Craftsmanship is exemplified by the retaining of control over the process of practice, especially at the production stage and involves the utilization of a high level of skill mastered for a social purpose (Alfoldy 2009; Frayling 2011; Sennett 2008).

Craft practices
Craft practices are those that display skills mastered and developed through practice, trial and error (Heidegger 1962; Johnson 1995; Polanyi 1974; Weiss 2008).

Culture
Culture is a collective term defined by the arts, customs, ideas and intellectual achievements of a society (Latour 1987; Frayling 2011).

Design
Design, says Flusser (1999), is so closely linked to technology and art that it is hard to contemplate one without the other two. Design can also form a plan or system for creating something (Rand 1947; Bierut 2012).

Digital
By combining and considering definitions from Barthes (1977), Crow (2006), Harris (2005) and McCullough (1996), it is suggested that digital can be explained as a paradigm in which each unit within a series is distinct. Digital is derived from the Latin *digitus* meaning a digit, such as a finger or toe (HarperCollins 1994); so is originally associated with human hands and feet within the physical environment.

Direct dye
Direct dye is another name for substantive dye and includes walnut husks or lichen (Dixon 1979; Hardman and Pinhey 2009).

Epistemology
Epistemology is the nature of the relationship we have with what we believe can be known (Bourdieu 1990; Crilly 2010).

Experiential knowledge
Experiential knowledge is the type of knowledge that is developed during the undertaking of an activity that adds to a person's experience and understanding of how something is done, such as creating in the studio (Nimkulrat 2007; Niedderer 2007).

Felt

Felt is one of the earliest man-made fabrics and pre-dates woven cloth. Wool fibers are rolled, agitated and pressed together so that they tangle to form a matted textile. The technique was first perfected by the nomadic tribes of Asia who relied on felt for clothing and shelter. They also attributed religious and cultural significance to it (Laufer 1930; Dixon 1979; Mullins 2009).

Fibers

Fibers are types of filaments or threads from which plant or animal tissue is formed. Wool fibers, for example, are composed of keratin, a protein, in complex arrangements. Different configurations of fiber structure result in the variety of characteristics seen in fleece from alternative breeds of sheep. Harvested fibers can be twisted or spun to create yarn using a spindle, spinning wheel or mechanical device (Dixon 1979; Fenner 1991; Franck 2005).

Founding

Founding is the process of casting a molten material, for example, metal or glass, into a prepared mold made of clay, sand or another metal, such as steel or brass (Cottingham 1824; Updike 1962).

Fused Deposition Modeling (FDM)

Fused Deposition Modeling (FDM) is an additive layering process that models CAD-simulated objects. Heated thermoplastic is squeezed from a nozzle and when it cools it sticks to the previous layer, thereby building up a solid structure (Delamore 2010; Hudson 2014).

Generic

Generic means belonging to a group or classification of something, rather than a specific example or brand type (Asher and Morreau 1995; Ollins 2008).

Ikat

Ikat is a type of resist dyeing process in which the warp and, or, the weft yarn of a weaving is colored before the textile is constructed. By tying the prepared ends or threads together tightly, then dyeing and untying, the resist technique creates a pattern that is characterized by a slight shifting of the edges of the image during the weaving process. The resist dyeing can be repeated multiple times before the textile is woven (Robinson 1969; Barnes 1989; Crill 1998; Riello 2010)

Intuitive knowledge

Intuitive knowledge explains Dewey (1934) is similar to a spark when a current flow of consciousness is suddenly joined with something old, and creates a subconscious realization about already knowing something (Foucault 1980; Niedderer 2004b).

Knowledge

Knowledge is more than simply a belief in something because it is objective (Latour 1987) and encompasses what is capable of being known. A shared knowledge also allows people to communicate more effectively (Polanyi 1974; Durling 2000).

Laser cutting

Laser cutting is a system that uses a CAD file to cut patterns into materials, such as wood, synthetic and natural fabrics. It is quick, is non-contact and when used for man-made materials it seals the edges where the laser cuts therefore avoiding fraying (Yusoff 2011; Mueller, Kruck and Baudisch 2013).

Lithography
Lithography is a technique that was developed in 1798 to duplicate images. An image is created from a greasy or waxy liquid plus crayon on an absorbent surface, originally limestone, then the surface is washed, inked and then paper laid on top and printed (Bankes 1976).

Luminosity
Luminosity is the intrinsic brightness of a color as perceived by the human eye (Townsend and Roberts 1999; Green and Kriss 2010).

Mordant
A mordant is a chemical required for natural dyeing with adjective dyes and can be added before, during or after dyeing, depending on the colors desired and the type of mordant being used. Different mordants produce a variety of shades. Tin also adds brightness to the final yarn, whereas iron tends to make the colors dull and chrome helps to increase softness (Dixon 1979; Hardman and Pinhey 2009).

Narrative
Narrative is a form of storytelling that uses sequencing. Culture depends on narrative to build up meaning, and allows people to interpret as well as construct memories (Genette 1983; Bal 1985).

Ontology
Ontology defines what can be known, its character and what we think or believe it can be (Disessa and Cobb 2004; Gray and Malins 2004; Pigrum 2007).

Palaeolithic
The Palaeolithic era marked the beginning of the use of hand tools and plant fibers by hominids, the family that includes modern humans and their fossil ancestors. The period coincided with the end of the last ice age, and is estimated to have lasted from 2.6 million until 12,000 years ago. The first cave art and elementary processing of pigment occurred during this time (Beltran 1999; Clottes 2010).

Pigment ink
Pigment ink is produced from non-organic particles that are inert and ground to a fine consistency so that once in a solution, they are able to travel freely through the print heads of an inkjet printer. Early pigments, however, were merely clay or earth and were mixed with animal fat or saliva to form a spreadable paste. Unlike organic dye, pigment ink is more colorfast although it tends to provide a less wide gamut (Delamare and Guineau 2000; Johnson 2007).

Post-processing
Post-processing refers to operations that are conducted after the production stage in textile manufacturing, such as fixing by steaming for reactive or acid dyes. Reasons can include improving the color gamut or fastness of the dyes (Ujiie 2006; Xin 2006).

Practice

Practice involves conducting studio inquiry creatively to produce works of art or design (Bolt 2006; Durling and Niedderer 2007; Sullivan 2008; Wallis 2010).

Practice-based research

Practice-based research can be defined as the use of creative practice aligned to the conventions of research and its associated research methods, methodologies, problematic and theoretical possibilities grounded in the artefact or its means of creation (Biggs 2002; Biggs and Büchler 2008; Candy and Edmonds 2011; Durling 2000; Nimkulrat 2009b).

Practice-led research

Practice-led research involves the use of the process of creative practice aligned to the conventions of research to lead the researcher to consider the nature of practice (AHRC 2007; Candy 2006; Nimkulrat 2007).

Print heads

The print heads of an inkjet printer or a 3D printer allow the printing material to be deposited or extruded from the printer onto a substrate (Campbell 2008; Hudson 2014).

Printing

From Benjamin (1936), Campbell (2008) and Collis and Wilson (2012), printing is the process of producing printed material and may include marking or the application of dye or ink to a surface. Printing comes from the Latin *premere*, via the French *periendre*; which means to press, the using of human hands or feet, to apply physical force (HarperCollins 1994; *Oxford English Dictionary* 1998).

Procedural knowledge

Procedural knowledge is the type of knowledge associated with the processes of how something is done. For example, it is used when making an artefact and involves the application of a series of tasks (Durling and Niedderer 2007; Niedderer 2009d).

Process

Process is a term that defines a set of procedures converting meaning from one form to another (Barrett 2007b; Latour 1987; Scrivener 2002a).

Rapid prototyping

Rapid prototyping is a form of additive three-dimensional printing that emerged in the late 1980s, primarily for the engineering industry. It allows designers to create models that are developed as concepts, and then produce them digitally (Chua, Leong and Lim 2010).

Rasterization

Rasterization is a process that takes a vector image and translates it into a pixelated version suitable for output on a digital printer. It is similar to the system originally commissioned by King Louis XIV in 1692, for use by the Imprimerie Royale, that interpreted early typeface designs as small squares prior to being produced as metal type (Bringhurst 1992; Lupton 2014).

Rapid prototyping

Rapid prototyping is a system in which layer upon layer of horizontal slices are printed and assembled, one on top of each other, and bonded together to form a 3D object (Chua, Leong and Lim 2010; Yusoff 2011).

Reactive dye

Reactive dye is a dye that is dissolved in a carrier liquid and tends to be more vibrant in color than pigment ink, although less stable (Christie 2001; Hoskins 2005).

Reproduction

Reproduction is a process or act that enables multiple copies of an original item to be made (Benjamin 1936; Bringhurst 1992; Sennett 2008).

Resist dyeing

Resist dyeing covers a range of techniques that create a barrier to dye being absorbed into a fabric, including waxing, binding, stitching or clamping. The earliest example was resist printing on textiles was discovered in Egypt (Collis and Wilson 2012).

Silk

Silk is a natural filament produced by a moth, such as the *Bombyx mori* that feeds on white mulberry leaves, to make a cocoon. A single cocoon contains around one kilometer of continuous thread, and its strength, warmth and luster has made it a popular fiber for textile production for thousands of years. Sericulture is the skill of rearing silk worms for raw silk (Fenner 1991; Aruga 1994).

Stamping

Stamping is the process of applying an engraved, or previously inked block, down onto a surface to create an impression or mark (Updike 1962; Laucy 2010).

Stencil

A stencil is a two-dimensional shape cut to form a negative template. The shape is placed over a substrate and paint or similar dyeing substance is applied through the cut-out area to produce an image on the surface below (Smith 1994; Gale and Kaur 2002)

Substantive dye

Substantive dye, or direct dye, is a category of dye that is sufficient on its own to fix color to natural fibers at a molecular level. Lichens were traditionally the most important source of substantive dyestuffs (Dixon 1979; Hardman and Pinhey 2009).

Substrate

Substrate is a term for a base structure or cloth, upon which a physical process such as printing is applied (Updike 1962; Braddock Clarke and O'Mahony 2007; Crow 2006).

Symbolism

Symbolism is the conveying of meaning through symbols. A symbol on its own does not supply new information, says Polanyi (1974), but it is its method of representation combined with how this is performed and understood by intelligent beings that articulates a specific expressive meaning (Whitehead 1985).

Synthetic dye
Synthetic dye is a man-made alternative to natural dye developed through chemical synthesis. In 1856 William Henry Perkin discovered mauveine, the first aniline dye (Christie 2001; Frayling 2011).

Tattoo
A tattoo is a mark or pattern created by repeatedly puncturing the surface of the human skin with a sharp implement, and applying pigment to the open areas in order to form a permanent design. The composition of the design contains symbolism that can reflect a person's cultural origins (Bradley 2000; DeMello 2000; Jablonski 2004).

Technology
Technology is the practical application of scientific knowledge (Heidegger 1977; Polanyi 1974; Latour 1987).

Textile
Using considerations sourced from Pittman and Townsend (2009), Quinn (2009), and Townsend, Briggs-Goode and Northall (2010), textile may be a material produced by crafting fibers together such as weaving, felting or knitting.

Theory
Theory involves a set of interrelated concepts that describe a phenomenon (Charmaz 1990; Friedman 2008; Glaser 1992; Glaser and Strauss 1967).

Tie-dye
Tie-dye, like batik, is a cloth resist dyeing technique. The constructed fabric, rather than the warp or weft threads of ikat, is tied tightly before dyeing to stop color penetrating into specific areas of the fabric. Once untied, the un-dyed areas of the cloth reveal a pattern formed by the tying process (Robinson 1969; Riello 2010).

Tool
A tool is an implement that is used to carry out a specific task, such as a lever that allows the human arm to extend its functionality (Flusser 1999; Sennett 2008).

Traditional
Traditional can be used to describe something that exists in the present, but has social origins in the past (Adamson 2009; Dewey 1910; Yanagi 1974).

Transferable knowledge
Transferable knowledge is capable of being communicated from one situation or person to another, such as that shared between a master and an apprentice when the latter picks up an understanding of how something is done that is not always voiced, overtly obvious or able to be written down (Sennett 2008; Nimkulrat 2007; Niedderer and Roworth-Stokes 2009).

Viscose

Viscose is a fiber produced from plants and trees, such as bamboo. It can be created chemically with caustic soda, also called sodium hydroxide, that is widely used for soap making (Waite 2009).

Woodblock

A woodblock is a piece of wood that can be carved or fashioned to allow prints to be made from its carved relief surface. After carving an image, the surface of the wood is inked up and pressure applied to print the design onto a substrate, such as paper or fabric. Most early woodblocks were the result of a division of labor and therefore did not accurately represent the original work of an artist, but rather a translation made by a cutter and printer (Chamberlain 1978; Uglow 2007; Riello 2010).

Wool

Wool is a natural fiber from an animal, such as a sheep or goat, that is extremely absorbent, retains heat and makes an efficient insulator (Cottle 2010). Its elasticity and durability make it a popular choice for textile production. Many varieties of sheep and goat are bred throughout the world, including in New Zealand, Australia, North America and the U.K., with the fleece ranging from the coarse Welsh Mountain, with a short staple length of between 5 and 10 centimeters, the Swaledale with a 20 to 30 centimeters staple length, to the lustrous Australian cashmere goat with a staple of around 10 centimeters (Dixon 1979; McGregor and Butler 2008).

Workmanship of certainty

Workmanship of certainty is the undertaking of practice in which the end result is predetermined at the outset (Marcus 2008; Pye 1968; Sontag 2008).

Workmanship of risk

Workmanship of risk is the conducting of practice in which the nature of the outcome is uncertain (Dormer 1990; Press 2007; Pye 1968).

BIBLIOGRAPHY

Ahmed, S. (2010), "Third of Adults 'Still Take Teddy Bear to Bed'," *Daily Telegraph*. Available online: www. telegraph.co.uk (accessed 22 February 2012).

AHRC. (2007), "Arts and Humanities Research Council: Practice-led Research." Available online: http:// www.ahrc.ac.uk (accessed 10 December 2011).

Alfoldy, S. (2009), "Craft, Space and Interior Design," American Craft, 68 (1): 74–6.

Althusser, L. ([1971] 2008), *On Ideology*, London and New York: New Left Books.

Aristotle ([1996] 2008), *Physics*, trans. R. Waterfield, Oxford: Oxford University Press.

Arnheim, R. (1969), *Visual Thinking*, Berkeley, CA: University of California Press.

Arthur, L. (2004), *Robert Stewart Design 1946–95*, New Brunswick, NJ: Rutgers University Press.

Aruga, H. (1994), *Principles of Sericulture*, Boca Raton, FL: CRC Press.

Asher, N. and M. Morreau (1995), "What Some Generic Sentences Mean," in G. Carlson (ed.), *The Generic Book*, 300–8, Chicago, IL, and London: University of Chicago Press.

Aubert, M. et al. (2014), "Pleistocene Cave Art from Sulawesi, Indonesia," *Nature*, 514: 223–7.

Bal, M. (1985), *Narratology: Introduction to the Theory of Narrative*, Toronto, ON, and London: University of Toronto Press.

Bankes, H. ([1813] 1976), *Henry Bankes's Treatise on Lithography*, London: Printing Historical Society.

Banks, J. et al. (1989). *Multicultural Education*, Needham Heights, MA: Allyn & Bacon.

Banksy. (2006), *Wall and Piece*, London: Century.

Barber, E. (1990), *Prehistoric Textiles: The Development of Cloth in the Neolithic and Bronze Ages, with Special Reference to the Aegean*, Princeton, NJ: Princeton University Press.

Barfield, N. and Quinn, M. (2004), "Research as a Mode of Construction: Engaging with the Artefact in Art and Design Research," *Working Papers in Art and Design*, 3. Available online: http://sitem.herts.ac.uk/ artdes_research/papers/wpades/vol3/nbfull.html (accessed 29 May 2010).

Barnes, R. (1989), *The Ikat Textiles of Lamalera: A Study of an Eastern Indonesian Weaving Tradition*, Leiden: E. J. Brill.

Barnes, R. (ed.) (1999), *Textiles in Indian Ocean Societies*, Oxford: Ashmolean Museum.

Barrett, E. (2007a), "Experiential Learning in Practice as Research: Context, Method, Knowledge," *Journal of Visual Arts Practice*, 2 (2): 115–24.

Barrett, E. (2007b), "Studio Enquiry and New Frontiers of Research," *Studies in Material Thinking*, 1 (1): 1–2.

Barthes, R. (1977), *Image, Music, Text*, London: HarperCollins.

Beddard, H. and D. Dodds (2009), *Digital Pioneers*, London: V&A.

Benjamin, W. (1936), *The Work of Art in the Age of Mechanical Reproduction*, trans. J. Underwood, London: Penguin.

Beckow, S. (1999), "Culture, History and Artifact," in T. Schlereth (ed.), Material Culture Studies in America: An Anthology Selected, Arranged, and Introduced by Thomas J. Schlereth, 114–23, Lanham, MD: Altamira Press.

Beltran, A. (1999), *The Cave of Altamira*, New York: Harry N. Abrams.

Berger, J. (1972), *Ways of Seeing*, London: Penguin.

Bide, M. (2007), "Environmentally Responsible Dye Application," in R. Christie (ed.), *Environmental Aspects of Textile Dyeing*, 74–92, Cambridge: Woodhead Publishing.

Bide, M. (2010), *Kirk-Othmer Encyclopedia of Chemical Technology*, London: John Wiley & Sons.

Bierut, M. (2012), *Seventy-Nine Short Essays on Design*, Princeton, NJ: Princeton Architectural Press.

Biggs, M. (2000), "Editorial: The Boundaries of Practice-based Research," Working Papers in Art Design, 1. Available online: http://www.herts.ac.uk/artdes_research/papers/wpades/vol1/intro.html (accessed 18 April 2011).

Biggs, M. (2002), "The Role of Artefact in Art and Design Research," *International Journal of Design, Sciences and Technology*, 10 (2): 19–24.

Biggs, M. and D. Büchler (2008), "Eight Criteria for Practice-based Research. The Creative and Cultural Industries," *Art, Design and Communication in Higher Education*, 7 (1): 5–18.

Bolt, B. (2006), "Materializing Pedagogies," Working Papers in Art and Design, 4. Available online: http://www.herts.ac.uk/artdes_researchpapers/wpades/vol4/bbfull.html (accessed 20 September 2010).

Bourdieu, P. (1990), *The Logic of Practice*, trans. R. Nice, Stanford, CA: Stanford University Press.

Bowles, M. (2013), "The People's Print." Available online: http://www.tfrc.org/research/the-peoples-print/ (accessed 28 May 2013).

Bowles, M. and C. Isaac (2009), *Digital Textile Design*, London: Laurence King.

Braddock Clarke, S. and M. O"Mahony (2007), *Techno Textiles 2: Revolutionary Fabrics for Fashion and Design*, London: Thames & Hudson.

Bradley, J. et al. (2000), *Written on the Body: The Tattoo in European and American History*, London: Reaktion.

Brandeis, S. (2003), "Post-digital Textiles: Re-discovering the Hand," Proceedings "Hands On" International Surface Design Conference, 3–8 June, Kansas.

Brandeis, S. (2007), "Academia Is Taking Digital Printing to New Heights." Available online: http://www.techexchange.com/thelibrary/academia.html (accessed 10 May 2011).

Briggs-Goode, A. (1997), "A Study of Photographic Images, Processes and Computer Aided Textile Design," PhD thesis, Nottingham Trent University.

Bringhurst, R. (1992), *The Elements of Typographic Style*, Point Roberts, WA: Hartley & Marks.

Britt, H., J. Stephen-Cran and E. Bremner (2013), "Awaken: Contemporary Fashion and Textile Interpretation of Archival Material," Futurescan 2: Collective Voices, 10–11 January, Sheffield Hallam University.

Brooks, H. (1967), "The Dilemmas of Engineering Education," *IEEE Spectrum*, 4 (2): 89–91.

Breuil, H. (1952), *Four Hundred Centuries of Cave Art*, trans. M. Boyle, Dordogne: Centre d"Etudes et de Documentation Prehistoriques.

Bryman, A. (2004), *Social Research Methods*, Oxford: Oxford University Press.

Büchler, D. (2006), "Contextualizing Perception in Design," *Working Papers in Art and Design*, 4. Available online: http://sitem.herts.ac.uk/artdes_research/papers/wpades/vol4/dbfullhtml (accessed 29 May 2010).

Buechley, L. et al. (2008), "The LilyPad Arduino: Using Computational Textiles to Investigate Engagement, Aesthetics and Diversity in Computer Science Education," in CHI '08 Proceedings of the SIGCHI Conference on Human Factors in Computing Systems, 5–10 April, Florence.

Bumpus, J. (2014), "There's Still Something about Mary," *Vogue*. Available online: www.vogue.co.uk/news/2014/10/23/mary-katrantzou-interview (accessed 15 February 2015).

Bunce, G. (1993), "An Investigation into the CAD/CAM Possibilities in the Printing of Textiles," PhD thesis, Nottingham Trent University.

Bunce, G. (2005), "Digital Print: To Repeat or Not to Repeat," Proceedings Creativity: Designer Meets Technology International Conference, 26–27 September, Copenhagen.

Bunnell, K. (1998), "The Integration of New Technology into Designer-Maker Practice," PhD thesis, The Robert Gordon University.

Bunnell, K. (2001), "Developing a Methodology for Practice-based Ceramic Design Research," Artelogi, 8: 9–17.

Bunnell, K. (2004), "Craft and Digital Technology," Proceedings World Crafts Council 40th Anniversary Conference, 1 June, Metsovo, Greece.

Calvino, I. (1997), *The Literature Machine: Essays*, trans. P. Creagh, London: Vintage.

Campbell, J. (2008), "Digital Printing of Textiles for Improved Apparel Production," in C. Fairhurst (ed.), *Advances in Apparel Production*, 222–49, Cambridge: Woodhead.

Campbell, J., H. Britt, A. Shaw and V. Begg (2008), *Mackintosh Re-interpreted*, Glasgow: Centre for Advanced Textiles, Glasgow School of Art.

Candy, L. (2006), "Practice Based Research: A Guide." Creativity and Cognition Studios. Available Online: http//:www.creativityandcognition.com (accessed 21 October 2012).

Candy, L. and E. Edmonds (eds) (2011), *Interacting: Art, Research and the Creative Practitioner*. Farringdon: Libri Publishing.

Carden, S. (2007), "Historical Consciousness and Mathematical Creativity: How the Most Important Formula in Mathematics Can be Deduced in Six Acts," Proceedings of the International Conference Reflections on Creativity: Exploring the Role of Theory in Creative Practices, Duncan of Jordanstone College of Art and Design.

Carden, S. (2011a), "Authenticity in Digital Printing," DPPI '11 Proceedings of the International Conference Designing Pleasurable Products & Interfaces, 40, ACM Digital Library, June.

Carden, S. (2011b), "Digitally Printed Textiles: A Number of Specific Issues Surrounding Research," *Journal of Craft Research*, 2 (1): 83–95.

Carden, S. (2011c), "A Critique of the Technologies Used in Digitally Printed Textile Design," CIPED Agenda for Design, International Conference, Fundação Gulbenkian, Lisbon, October.

Carden, S. (2012), "Digitally Printed Textiles: New Processes and Theories," Design Research Society, Chulalongkorn University, Bangkok, July. DRS, 2: 199–212.

Carden, S. (2013a), "Craft Informed Digital Textile Printing," Proceedings International Conference Crafts as Change Makers in Sustainably Aware Futures, Plymouth University, September. *Making Futures*, 2: (1): 272–81.

Carden, S. (2013b), "Innovative Synthesis of Craft and Digital Processes: Theory Building through Textile Design Practice," PhD thesis, The Glasgow School of Art.

Carden, S. (2015), "Processes within Digitally Printed Textile Design," in F. Kane, N. Nimkulrat and K. Walton (eds), *Crafting Textiles in the Digital Age*, London: Bloomsbury.

Carlisle, H. (2002), "Towards a New Design Strategy: A Visual and Cultural Analysis of Small-Scale Pattern on Clothing," PhD thesis, Nottingham Trent University.

Casey, B. (2012), "Foreword," in *Type Matters!*, London and New York: Merrell.

Chamberlain, W. (1978). *The Thames & Hudson Manual of Woodcut Printing and Related Techniques*, London: Thames & Hudson.

Chang, A. (2013), "Angel Chang." Available online: http://:www.angelchang.com (accessed 12 March 2013).

Charmaz, K. (1990), "'Discovering' Chronic Illness: Using Grounded Theory," *Social Science & Medicine*, 30 (11): 1161–72.

Christie, R. (2001), *Colour Chemistry*, London: Royal Society of Chemistry.

Chua, C., K. Leong and C. Lim (2010), *Rapid Prototyping: Principles and Applications*, Singapore: World Scientific.

Cline, E. (2012), *The Oxford Handbook of the Bronze Age Aegean*, Oxford: Oxford University Press.

Clottes, J. (2003a), *Chauvet Cave: The Art of Earliest Times*, trans. P. Bahn, Utah, UH: University of Utah Press.

Clottes, J. (2003b), *Return to the Chauvet Caves, Excavating the Birthplace of Art*: The First Full Report, London: Thames & Hudson.

Clottes, J. (2010), *Cave Art*, London and New York: Phaidon Press.

Collis, A. and J. Wilson (2012), "Colour Accuracy in Digitally Printed Textiles: What You See is Not (Always) What You Get," *Journal of the International Colour Association*, 9: 20–31.

Cooper-Hewitt Museum (1987), *Printed Textiles 1760–1860 in the Collection of the Cooper-Hewitt Museum*, Washington, DC: Smithsonian Institute.

Cottingham, L. 1824. *The Smith Founder and Ornamental Metal Maker's Director*, 3rd edn, London: J. Brooks.

Cottle, D. (ed.) (2010), *The International Sheep and Wool Handbook*, 2nd edn, Nottingham: Nottingham University Press.

Crill, R. (1998), *Indian Ikat Textiles*, London: V&A Publishing.

Crill, R. (2008), *Chintz: Indian Textiles for the West*, London: V&A Publishing.

Crilly, N. (2010), "The Structure of Design Revolutions: Kuhnian Paradigm Shifts in Creative Problem Solving," *Design Issues*, 26 (1): 54–66.

Crow, D. (2006), *Left to Right: The Cultural Shift from Words to Pictures*, Lausanne: AVA Academia.

Crump, S. (1992), *Apparatus and Method for Creating Three-Dimensional Objects*, US Patent 5121329 A.

Daniels, H. (2001), *Vygotsky and Pedagogy*, London: Routledge.

Delamare, F. and B. Guineau (2000), *Colour: Making and Using Dyes and Pigments*, London: Thames & Hudson.

Delamore, P. (2004), "3D Printed Textiles and Personalised Clothing." Available online: http://www.academia.edu/917613/3D_Printed_Textiles_and_Personalised_Clothing (accessed 11 August 2012).

Delamore, P. (2010), "Considerate Design for Personalised Fashion," in T. Inns (ed.), *Designing for the 21st Century: Interdisciplinary Methods and Findings*, 67–86, Farnham: Gower.

Deleuze, P. and F. Guattari (1996), *What is Philosophy?*, Columbia, NY: Columbia University Press.

DeMello, M. (2000), *Bodies of Inscription: A Cultural History of the Modern Tattoo Community*, Durham, NC: Duke University Press.

Dewey, J. (1910), *How We Think*, Lexington, MA: D. C. Heath.

Dewey, J. (1934), *Art as Experience*, New York: Perigee Books.

Disney Research (2014), "Teddy Bears at the Push of a Button: CMU-Disney Researcher Invents 3D Printing Technique for Making Soft, Cuddly Stuff." Available online: www.disneyresearch.com (accessed 30 October 2014).

Disessa, A. and P. Cobb (2004), "Ontological Innovation and the Role of Theory in Design Experiments," *Journal of the Learning Sciences*, 13 (1): 77–103.

Dixon, M. (1979), *The Wool Book*, London & New York: Hamlyn.

Dormer, P. (1990), *The Meaning of Modern Design*, London: Thames & Hudson.

Dormer, P. (ed.) (1997), *The Culture of Craft: Status and Future Studies in Design and Material Culture*, Manchester: Manchester University Press.

Douglas, J. (1985), *Creative Interviewing*, London: Sage.

Dreyfus, H. (1972), *What Computers Still Can't Do: A Critique of Artificial Reason*, Cambridge, MA: MIT Press.

Durling, D. (2000), "Reliable Knowledge in Design," Working Papers in Art and Design, 1. Available online: http://www.herts.ac.uk/artdes_research/papers/wpades/vol1/ddfull.html (accessed 10 November 2011).

Durling, D. and K. Niedderer (2007), "The Role and Use of Creative Practices in Research and its Contribution to Knowledge. Emerging Trends in Design Research," Proceedings IASDR International Conference, 12–15 November, Hong Kong.

Dutton, L. (1992), "Cochineal: A Bright Red Animal Dye," MSc Thesis, Baylor University, Waco, Texas.

Earley, R. (2009), "Ethical Fashion Forum." Available online: http://ethicalfashionforum.com/source-directory/member/90 (accessed 27 September 2011).

Eckert, C., P. Delamore and C. Bell (2010), "Dialogue across Design Domains: Rapid Prototyping in Aerospace and Fashion," Proceedings 11th International Design Conference, 17–20 May, Dubrovnik.

Eco, U. (1989), *The Open Work*, trans. A. Cancogni, Cambridge, MA: Harvard University Press.

Edwards, (2007), "Cloth and Community: The Local Trade in Resist Dyed and Block-printed Textiles in Kachchh District, Gujarat," *Textile History*, 38 (2): 179–97.

Einstein, A. (1987), *Collected Papers of Albert Einstein, Volume 1: The Early Years, 1879–1902*, Princeton, NJ: Princeton University Press.

Erlandson, D. et al. (1993), *Doing Naturalistic Inquiry: A Guide to Methods*, London and New Delhi: SAGE.

Fenner, M. (1991), *Keeping Silkworms*, Melbourne: Penguin.

Finlay, L. (2006), "The Body's Disclosure in Phenomenological Research," *Qualitative Research in Psychology*, 3 (1): 19–30.

Flusser, V. (1983), *Towards a Philosophy of Photography*, trans. A. Matthews, London: Reaktion.

Flusser, V. (1999), *The Shape of Things: A Philosophy of Design*, trans. A. Matthews, London: Reaktion.

Fontana, A. and J. Frey (2005), "The Interview: From Neutral Stance to Political Involvement," in N. Denzin and Y. Lincoln (eds), *Handbook of Qualitative Research*, 3rd edn, 695–728, Thousand Oaks, CA: SAGE.

Foucault, M. (1970), *The Order of Things: An Archaeology of the Human Sciences*, London and New York: Routledge Classics.

Foucault, M. (1980), *Power/Knowledge: Selected Interviews and Other Writings 1972–1977*, trans. C. Gordon et al., New York: The Harvester Press.

Fox, S. (2013), "Shelly Fox." Available online: http://shellyfox.com (accessed 24 March 2013).

Franck, R. (ed.) (2005), *Bast and Other Plant Fibers*, Cambridge: Woodhead Publishing.

Frayling, C. (1993), "Research in Art and Design," *Royal College of Art Research Papers*, 1 (1): 1–5.

Frayling, C. (2011), On Craftsmanship: Towards a New Bauhaus, London: Oberon Masters.

Freeland, C. (2001), *But is it Art?*, Oxford: Oxford University Press.

Friedman, K. (2008), "Research into, by and for Design," *Journal of Visual Arts Practice*, 7 (2): 153–60.

Fu, Z. (2006), "Pigmented Ink Formulation for Digital Textile Printing," in H. Ujiie (ed.), *Digital Printing of Textiles*, 218–32, Cambridge: Woodhead Publishing.

Fuad-Luke, A. (2002), *The Eco-Design Handbook*, London: Thames & Hudson.

Gale, C. and J. Kaur (2002). *The Textile Book*, Oxford: Berg.

Gardner, H. (2006), *Multiple Intelligences*: New Horizons, New York: Basic Books.

Genette, G. (1983), *Narrative Discourse: An Essay in Method*, trans. J. Lewin, New York: Cornell University Press.

Gerber, A. (2004), *All Messed Up: Unpredictable Graphics*, London: Laurence King.

Glaser, B. (1992), *Emergence vs Forcing: Basics of Grounded Theory Analysis*, Mill Valley, CA: Sociology Press.

Glaser, B. and A. Strauss (1967), *The Discovery of Grounded Theory: Strategies for Qualitative Research*, New York: Aldine.

Goldsworthy, K. (2009), "Resurfaced: Using Laser Technology to Create Innovative Surface Finishes," Cutting Edge: Lasers and Creativity Symposium, 4 November, Loughborough.

Goldsworthy, K. (2013a), "Laser-finishing: A New Process for Designing Recyclability in Synthetic Textiles," PhD thesis, Central St. Martins, London.

Goldsworthy, K. (2013b), "Design for Cyclability." Available online: http://www.tfrc.org.uk (accessed 28 May 2013).

Gray, C. and J. Malins (2004), *Visualizing Research: A Guide to the Research Process in Art and Design*, Farnham and Burlington, VT: Ashgate.

Green, P. and M. Kriss (2010), *Colour Management: Understanding and Using ICC Profiles*, Chichester: John Wiley & Sons.

Grossman, B. (2013), "Bathsheba Sculpture." Available online: http://www.bathsheba.com (accessed 21 September 2013).

Guba, E. and Y. Lincoln (1985), *Naturalistic Inquiry*, Thousand Oaks, CA: Sage.

Gubrium, J. and J. Holstein (1998), "Narrative Practice and the Coherence of Personal Stories," *The Sociological Quarterly*, 39 (1): 163–87.

Habermas, T. and C. Paha (2002), "Souvenirs and Other Personal Objects: Reminding of Past Events and Significant Others in the Transition to University," in J. D. Webster and B. K. Haight (eds), *Critical Advances in Reminiscence Work*, 123–38, New York: Springer.

Hall, I. and D. Hall (1996), *Practical Social Research: Project Work in the Community*, London: Macmillan.

Harari, Y. (2014), *Sapiens: A Brief History of Humankind*, London: Harvill Secker.

Hardman, J. and S. Pinhey (2009), *Natural Dyes, Marlborough*: The Crowood Press.

Harris, J. (2005), "Crafting Computer Graphics – A Convergence of Traditional and 'New' Media," *The Journal of Cloth and Culture*, 3 (1): 20–35.

Harrod, T. (2007), "Modernity and the Crafts," in S. Alfoldy (ed.), *NeoCraft*, 225–41, Halifax, NS: The Press of the Nova Scotia College of Art and Design.

Harthan, J. (1983), "The Development of Bookbinding Design," in P. Winckler (ed.), *Reader in the History of Books and Printing*, 102–12, Englewood, CO: Greenwood Press.

Heidegger, M. (1962), *Being and Time*, trans. J. MacQuarrie and E. Robinson, Malden, MA: Blackwell.

Heidegger, M. (1977), *The Question Concerning Technology and Other Essays*, trans. W. Lovitt, New York: HarperPerennial.

Henshilwood, C. et al. (2011), "A 100,000-Year-Old Ochre Processing Workshop at Blombos Cave, South Africa," *Science*, 334 (6053): 219–22.

Herbert, S. (1980), *Introduction to Printing: Craft of Letterpress*, London: Faber & Faber.

Herodotus. (1998), *The Histories*, trans. R. Waterfield, Oxford: Oxford University Press.

Hermitage. (2013), "Tattooed Body of a Tribal Chief." Available online: http://www.hermitagemuseum.org.html (accessed 4 February 2013).

Heyenga, L. and R. Ryan (2011), *Paper Cutting, Contemporary Artists: Timeless Craft*, San Francisco, CA: Chronicle Books.

Hohl, M. (2006), "This is Not Here: Connectedness, Remote Experiences and Telematic Art," PhD thesis, Sheffield Hallam University.

Hohl, M. (2009), "Designing the Art Experience: Using Grounded Theory to Develop a Model of Participants' Participation of an Immersive Telematic Artwork," *Digital Creativity*, 20 (3): 187–96.

Holler, A. and M. Gotz (2013), "Not Without My Teddy Bear: The Companions of Childhood," A Collaborative study produced by the Chances for Children through Play Foundation and the International Central Institute for Youth and Educational Television (IZI). Available online: www.br_online.de (accessed 30 June 2014).

Hopper, A. (2010), "Cambridge Ideas Change the World." Available online: http://www.cambridgenetwork.co.uk.html (accessed 12 June 2010).

Hoskins, S. (2005), *Inks*, London: A. & C. Black.

Huang, M. et al. (2011), "Human Cortical Activity Evoked by the Assignment of Authenticity when Viewing Works of Art," *Frontiers in Human Neuroscience*, 5 (134). Available online: http://journal.frontiersin.org/article/10.3389/fnhum.2011.00134/full#h10 (accessed 12 May 2013).

Hudson, S. (2014), "Printing Teddy Bears: A Technique for 3D Printing of Soft Interactive Objects," in CHI '14 Proceedings of the SIGCHI Conference on Human Factors in Computing Systems, 26 April – 1 May, Toronto, 459–68.

Hughes, B. (1999), *Dust or Magic*, Boston, MA: Addison-Wesley.

Hyman, J. (2006), *The Objective Eye: Color, Form, and Reality in the Theory of Art*, Chicago, IL: University of Chicago Press.

Indianetzone (2013), "Hand Printing and Dyeing." Available online: (accessed 30 April 2013).

Jablonski, N. (2004), "The Evolution of Human Skin and Skin Colour," *Annual Review of Anthropology*, 33: 585–623.

Jezequal. (2011), "Characterization and Origin of Black and Red Magdalenian Pigments from Grottes de la Garoma (Vallee Maoyenne de la Cause – France): A Mineralogical and Geological Approach of the Prehistorical Paintings," *Journal of Archaeological Science*, 38 (6): 1165–72.

Johnson, B. (2007), "Dye versus Pigment Ink Printers." Available online: http://www.earthboundlight.com (accessed 14 May 2013).

Johnson, P. (1995), "Naming the Parts," *Crafts*, 143: 42–5.

Johnson, P. (2003), *Art: A New History*, London: Weidenfeld & Nicolson.

Joseph, F. and C. Cie (2009), "Re-distributed Thinking: Paradigmatic Shifts in Textile Design Technologies and Methodologies." Available online: http://www.academia.edu/Documents/in/Digital_printing (accessed 7 April 2013).

Katrantzou, M. (2012), "Mary Katrantzou x Lesage." Available online: www.marykatrantzou.com/projects/mary-katrantzou-x-lesage/autumn-winter-(2012), (accessed 12 February 2015).

Katrantzou, M. (2014), "Mary Katrantzou x Lesage." Available online: www.marykatrantzou.com/projects/mary-katrantzou-x-lesage/spring-summer-2014 (accessed 12 February 2015).

Kawai, H. et al. (2002), "Longitudinal Study of Procedural Memory in patients with Alzheimer-type Dementia," *Brain and Nerve*, 54 (4): 307–11.

Keeling, R. (1981), "Ink Jet Printing," *Physics in Technology*, 12: 196–203.

Kerlogue, F. (2014), *Batik: Design, Style and History*, London: Thames & Hudson.

Kleiner, F. (2011), *Gardner's Art through the Ages: A Global History*, 14th edn, Boston, MA: Wadsworth.

Keats, D. (2000), *Interviewing: A Practical Guide for Students and Professionals*, Maidenhead: Open University Press.

Krogh, A. (2013), "Astrid Krogh." Available online: http://astridkrogh.com (accessed 15 March 2013).

Kyttanen, J. (2010), "Freedom of Creation." Available online: http://www.freedomofcreation.com (accessed 10 October 2011).

Langdridge, D. (2007), *Phenomenological Psychology: Theory, Research and Method*, Harlow: Pearson.

Latour, B. (1987), *Science in Action*, Cambridge, MA: Harvard University Press.

Laucy, J. (2010), *Fabric Stamping Handbook*, Concord, CA: C&T Publishing.

Laufer, B. (1930), "The Early History of Felt," *American Anthropologist*, 32 (1): 1–18.

Lawson, A. (2012), *Painted Canvas: Palaeolithic Rock Art in Western Europe*, Oxford: Oxford University Press.

Leach, B. (1976), *Culture and Communication: The Logic by Which Symbols are Connected*, Cambridge: Cambridge University Press.

Leaper, C. (2014), "Mary Katrantzou on Why She's Turned her Back on Digital Print." Available online: www.marieclaire.co.uk/blogs/547041/aw14-mary-katrantzou-shares-her-collection-inspiration (accessed 15 February 2015).

Lees-Maffei, G. and L. Sandino (2004), "Dangerous Liaisons: Relationships between Design, Craft and Art," *Journal of Design History*, 17 (3): 207–15.

Lemire, B. (2011), *Cotton*, Oxford: Berg.

Lupton, E. (2014), *Type on Screen: A Critical Guide for Designers, Writers, Developers, and Students*, New York: Princeton Architectural Press.

Luquiens, H. (1928), *Copper Plate Printing*, Hawaii: Honolulu Academy of Arts.

McCullough, M. (1996), *Abstracting Craft:* the Practiced Digital Hand, Cambridge, MA: MIT.

McDonough, W. and M. Braungart (2002), *Cradle to Cradle: Remaking the Way We Make Things*, New York: North Point Press.

McGregor, B. and K. Butler (2008), "Cashmere Production and Fleece Attributes Associated with Farm of Origin, Age and Sex of Goat in Australia," *Australian Journal of Experimental Agriculture*, 48 (8): 1090–98.

McLuhan, M. (1964), *Understanding Media*, London: Routledge.

McLuhan, M. and Q. Fiore (1967), *The Medium is the Message*, London: Penguin.

McLuhan, M. and B. Powers (1989), *The Global Village: Transformations in World Life and Media in the 21st Century*, New York and Oxford: Oxford University Press.

Manovich, L. (2001), *The Language of New Media*, Cambridge, MA: MIT Press.

Marcus, G. (2008), "Disavowing Craft at the Bauhaus: Hiding the Hand to Suggest Machine Manufacture," *The Journal of Modern Craft*, 1 (3): 345–56.

Makela, M. (2009), "The Place and the Product(s) of Making in Practice-led Research," in N. Nimkulrat and T. O"Liley (eds), *Reflections and Connections: On the Relationship Between Creative Production and Academic Research*, 29–38, Helsinki: UIAH.

Margolin, V. (2010), "Doctoral Education in Design: Problems and Prospects," *Design Issues*, 26 (3): 70–8.

Marshall, J. (1999), "The Role and Significance of CAD/CAM Technologies in Craft and Designer Maker Practice: With a Focus on Architectural Ceramics," PhD Thesis, University of Wales Institute, Cardiff.

Marx, K. (1976), *Capital, Volume 1: A Critique of Political Economy*, trans. B. Fowkes, London: Pelican.

Mahon, B. (2004), *The Man Who Changed Everything: The Life of James Clerk Maxwell*, Chichester: John Wiley & Sons.

Marzke, (1997), "Precision Grips Hand Morphology and Tools," *American Journal of Physical Anthropology*, 102 (1): 91–110.

Mehmet, S., L. Xiling and P. Khaitovich (2013), "Human Brain Evolution: Transcripts, Metabolites and their Regulators," *Nature Reviews Neuroscience*, 14: 112–27.

Mehta, R. (2013), "Occupational Hazards Caused in Textile Printing Operations." Available online: http://www.fiber2fashion.com (accessed 14 April 2013).

Mellars, P. (1996), *The Neanderthal Legacy: An Archaeological Perspective from Western Europe*, Princeton, NJ: Princeton University Press.

Menkes, S. (2012), "The Queen of Prints." Available online: www.nytimes.com/2012/02/2012/fashion/during-london-week-mary-katrantzou-talks-about-prints (accessed 10 February 2015).

Merleau-Ponty, M. (1960), "Eye and Mind," trans. C. Smith, in *Merleau-Ponty's Essays on Painting*, Evanston, IL: Northwestern University Press.

Merleau-Ponty, M. (2002), *Phenomenology of Perception*, trans. C. Smith, London and New York: Routledge.

Mirzoeff, N. (1999), *An Introduction to Visual Culture*, New York: Routledge.

Moser, L (2003), "ITMA 2003 Review: Textile Printing," *Journal of Textile and Apparel, Technology and Manufacturing*, 3 (3): 1–15.

Mueller, S., B. Kruck and P. Baudisch (2013), "LaserOrigami: Laser-Cutting 3D Objects," in Proceedings of the SIGCHI Conference on Human Factors in Computing Systems, 27 April – 2 May, Paris, 2585–92.

Mullins, W. (2009), *Felt*, Oxford and New York: Berg.

Natural History Museum (2013), "Seeds of Trade." Available online: http://www.nhm.ac.uk (accessed 14 February 2013).

Nicol, K. (2009), "Interview," in *The Royal College of Art*. Available online: http://www.rca.ac.uk (accessed 8 May 2010).

Niedderer, K. (2004a), "Why is There the Need for Explanation? – Objects and Their Realities," *Working Papers in Art and Design*, 3. Available online: http://sitem.herts.ac.uk/artdes_research/papers/wpades/vol3/knfull.html (accessed 10 April 2011).

Niedderer, K. (2004b), "Designing the Performative Object: A Study in Designing Mindful Interaction through Artefacts," PhD thesis, University of Plymouth.

Niedderer, K. (2007), "Mapping the Meaning of Experiential Knowledge in Research," *Design Research Quarterly*, 2 (1). Available online: http://www.drsq.org/issues/drsq2-2.pdf (accessed 15 March 2010).

Niedderer, K. (2009a), "Understanding Methods: Mapping the Flow of Methods, Knowledge and Rigour in Design Research Methodology," Proceedings IASDR International Conference, 18–22 October, Seoul, South Korea.

Niedderer, K. (2009b), "The Culture and Politics of Knowledge in Design Research: How to Develop

Discipline Specific Methodologies," Proceedings EKSIG International Conference, 19 June, Spokane, Washington, 1–18.

Niedderer, K. (2009c), "Sustainability of Craft as a Discipline?" Proceedings Making Futures Conference 1, 17–19 September, Plymouth, 165–74.

Niedderer, K. (2009d), "Relating the Production of Artefacts and the Production of Knowledge in Research," in N. Nimkulrat and T. O"Riley (eds), *Reflections and Connections: On the Relationship Between Creative Production and Academic Research*, 59–67, Helsinki: UIAH.

Niedderer, K. and R. Roworth-Stokes (2009), "The Role and Use of Creative Practice in Research and its Contribution to Knowledge," Proceedings IASDR International Conference, 18–22 October, Seoul, South Korea.

Nimkulrat, N. (2007), "The Role of Documentation in Practice-led Research," *Journal of Research Practice*, 3 (1): 1–8.

Nimkulrat, N. (2009a), "Creation of Artifacts as a Vehicle for Design Research," Proceedings Nordes International Conference. Available online: http://www.nordes.org (accessed 5 October 2010).

Nimkulrat, N. (2009b), "Paperness: Expressive Material in Textile Art from an Artist's Viewpoint," PhD thesis, Helsinki University of Art and Design.

Nimkulrat, N. (2010), "Material Inspiration: From Practice-led Research to Craft Art Education," *Journal of Craft Research*, 1 (1): 63–84.

Nonaka, I. and H. Takeuchi (1995), "Creating Knowledge in Practice," in *The Knowledge Creating Company: How Japanese Companies Create the Dynamics of Innovation*, 95–123, Oxford: Oxford University Press.

North, J. (1969), Mid-Nineteenth Century Scientists, London: Pergamon Press.

O"Connor, E. (2005), "Embodied Knowledge: The Experience of Meaning and the Struggle towards Proficiency in Glassblowing," *Ethnography*, 6 (2): 183–204.

Ollins, W. (2008), *The Brand Handbook*, London: Thames & Hudson.

Oppenheim, B. (1992), *Questionnaire Design, Interviewing and Attitude Measurement*, London: Continuum.

Ordoñez, M. (2000), "Island Quilts: Technology Reflected," in L. Welters and M. Ordoñez (eds), *Down by the Old Mill Stream: Quilts in Rhode Island*, 122–6, Kent, OH: Kent State University Press.

Ospitali, F., D. Smith and M. Lorblanchet (2006), "Preliminary Investigations by Raman Microscopy of Prehistoric Pigments in the Wall-Painted Cave of Roucadour, Quercy, France," *Journal of Raman Spectroscopy*, 37 (10): 1063–71.

Oxford English Dictionary (1998), Oxford: Oxford University Press.

Parry, L. (2005), *Textiles of the Arts and Crafts Movement*, London: Thames & Hudson.

Parry, L. (2009), *V&A Pattern:* William Morris, London: V&A Publishing.

Pasupathi, M. (2001), "The Social Constructions of the Personal Past and its Implications for Adult Development," *Psychological Bulletin*, 127 (5): 651–72.

Peirce, C. (1878), "How to Make Our Ideas Clear," *Popular Science Monthly*, 12: 286–302.

Peña, A. (2010), "The Dreyfus Model of Clinical Problem-Solving Skills Acquisition: A Critical Perspective," *Medical Education Online*, 15 (10). Available online: http://www.ncbi.nlm.nih.gov (accessed 6 September 2013).

Perner-Wilson, H. and L. Buechley (2010), "Making Textile Sensors from Scratch," in TEI '10 Proceedings of the Fourth International Conference on Tangible, Embedded, and Embodied Interaction, 25–7 January, Cambridge, MA, 349–52.

Perez-Seoane, M. (1999), *The Cave of Altamira*, New York: Harry N. Abrams.

Philpott, R. (2011), "Structured Textiles: Adaptable Form and Surface in Three Dimensions," PhD thesis, Royal College of Art.

Philpott, R. (2012), "Crafting Innovation: The Intersection of Craft and Technology in the Production of Contemporary Textiles," *Journal of Craft Research*, 3: 53–73.

Phillipson, M. (1995), "Managing 'Tradition': The Plight of Aesthetic Practices and their Analysis in a Technoscientific Culture," in C. Jenks (ed.), *Visual Culture*, 202–27, London: Routledge.

Pigrum, D. (2007), "The 'Ontology' of the Artist's Studio as Workplace: Researching Artist's Studio and the Art/Design Classroom," *Research in Post-Compulsory Education*, 12 (3): 291–307.

Pike, A. et al. (2012), "U-Series Dating of Palaeolithic Art in 11 Caves in Spain," *Science*, 336 (6087): 1409–13.

Pirsig, R. (1974), *Zen and the Art of Motorcycle Maintenance*, London: The Bodley Head.

Pittman, L. and K. Townsend (2009), "Designer/Makers are Key to Sustainable Textile Development," Proceedings Making Futures Conference 1, 17–19 September, Plymouth, 175–85.

Plato. (2000), *The Republic*, trans. T. Griffith, Cambridge: Cambridge University Press.

Polanyi, M. (1969), *Knowing and Being*, Chicago, IL: Chicago University Press.

Polanyi, M. (1974), *Personal Knowledge: Towards a Post-Critical Philosophy*, Chicago, IL: Chicago University Press.

Pollock, G. (2008), "An Engaged Contribution to Thinking About Interpretation in Research in/into Practice," *Working Papers in Art and Design*, 5. Available online: http://sitem.herts.ac.uk/artdes_research/papers/wpades/vol5/gpfull.html (accessed 10 September 2010).

Polo, M. (1997), *The Travels of Marco Polo*, trans. W. Marsden, Ware: Wordsworth Classics.

Popper, K. (1991), *Conjectures and Refutations: The Growth of Scientific Knowledge*, London and New York: Routledge.

Prakesh, O. (2012), "Indian Textiles in the Indian Ocean Trade in the Early Modern Period." Available online: http://www2.lse.ac.uk/ecenomicHistory/Research/GEHN/HELSINKIPrakesh.pdf (accessed 15 February 2013).

Press, M. (2007), "Handmade Futures: The Emerging Role of Craft Knowledge in Our Digital Culture," in S. Alfoldy (ed.), *NeoCraft*, 251–68, Halifax, NS: The Press of the Nova Scotia College of Art and Design.

Provost, J. (2008), "Effluent Improvement by Source Reduction of Chemicals used in Textile Printing," *Journal of The Society of Dyers and Colourists*, 108 (5): 260–4.

Pye, D. (1968), *The Nature and Art of Workmanship*, London: Fox Chapel Publishing.

Pye, D. (1978), *The Nature and Aesthetics of Design*, London: Cambium Press.

Quek, S. (2015), "Mary Katrantzou AW15." Available online: www.dazeddigital.com/fashion/article/23747/1/mary-katrantzou-livestream (accessed 10 February 2015).

Quinn, B. (2009), *Textile Designers: At the Cutting Edge*, London: Laurence King.

Quinn, B. (2011), *Textile Futures*, Oxford: Berg.

Quinn, B. (2012), *Fashion Futures*, London: Merrell.

Rancière, J. (2004), *The Politics of Aesthetics: The Distribution of the Sensible*, trans. G. Rockhill, New York and London: Continuum.

Rand, P. (1947), *Thoughts on Design*, New York: Wittenborn Schultz.

RapidToday. (2013), "Rapid Protyping Struggles to Find Niche in Art," Design. Available online: http://rapidtoday.com (accessed 30 October 2013).

Rehbein, M. (2010), *Digital Textile Printing and the Influence on Design*. GRIN Verlag.

Riello, G. (2010), "Asian Knowledge and the Development of Calico Printing in Europe in the Seventeenth and Eighteenth Centuries," *Journal of Global History*, 5: 1–28.

Riello, G. (2013), *Cotton: The Fabric that Made the Modern World*, New York: Cambridge University Press.

Rigley, S. (2004), "Re-tooling the Culture for an Empire of Signs," *Eye: The International Review of Graphic Design*, 52: 56–64.

Rigley, S. (2013), "Digital Kanthas: Print Waste or Ornament," *Journal of Craft Research*, 4 (2): 245–64.

Robertson, F. (2013), *Print Culture*, New York: Routledge.

Robinson, S. (1969), *A History of Printed Textiles*, Cambridge, MA: MIT Press.

Robson, C. (1993), *Real World Research: A Resource for Social Scientists and Practitioner Researchers*, Oxford: Blackwell.

Roebroeks, W. et al. (2012), "Use of Red Ochre by Neandertals," PNAS, 109 (6): 1889–94.

Roy, M. (1978), "Dyes in Ancient and Medieval India," *Indian Journal of History of Science*, 13 (2): 83–113.

Russell, A. (2009), "Alex Russell," in B. Quinn (ed.), *Textile Designers: At the Cutting Edge*, 12–19, London: Laurence King.

Russell, A. (2011), *The Fundamentals of Printed Textile Design*, Worthing: AVA Publishing.

Russell, A. (2013), "Repeatless: The Use of Digital Technology to Extend the Possibilities of Printed Textile Design," in *Proceedings of the 1st International Conference on Digital Technologies for the Textile Industries*, 5–6 September, University of Manchester.

Sanders, E. (2008), "The Centre for Advanced Textiles: A Case Study in UK Digital Textile Culture," in *Proceedings Textile Society of America Symposium*, 24–27 September.

Sardar, M. (2000), "Europe and the Islamic World, 1600–1800," in *Heilbrunn Timeline of Art History*, New York: The Metropolitan Museum of Art.

Scagnetti, G. (2012), "Emergent Design Landscape." Available online: http://drs2012bangkok.org/index.php?page=Design-Exhibition (accessed 10 January 2013).

Schoeser, M. (2004), "Textiles," in L. Arthur (ed.), *Robert Stewart Design 1946–95*, 67–89, New Brunswick, NJ: Rutgers University Press.

Schön, D. (1983), *The Reflective Practitioner: How Professionals Think in Action*, New York: Basic Books.

Schön, D. (1987), *Educating the Reflective Practitioner*, San Francisco, CA: Jossey-Bass.

Scrivener, S. (2002a), "Characterising Creative-Production Doctoral Projects in Art and Design," International *Journal of Design Sciences and Technology*, 10 (2): 25–44.

Scrivener, S. (2002b), "The Art Object Does Not Embody a Form of Knowledge," *Working Papers in Art and Design*, 2. Available online: http://www.herts.ac.uk/artdes_research/papers/wpades/vol3/ssfull.html (accessed 29 May 2010).

Scrivener, S. (2009), "The Roles of Art and Design Process and Object in Research," in Nimkulrat, N. and T. O"Riley (eds), *Reflections and Connections: On the Relationship Between Creative Production and Academic Research*, 69–80, Helsinki: UIAH.

Scrivener, S. and P. Chapman (2004), "The Practical Implications of Applying a Theory of Practice Based Research: A Case Study," *Working Papers in Art and Design*, 3. Available online: http://sitem.herts.ac.uk/artdes_research/papers/wpades/vol3/ssfull.html (accessed 11 June 2010).

Sekaran, U. (1992), *Research Methods for Business: A Skill Building Approach*, New York: John Wiley & Sons.

Sennett, R. (2008), *The Craftsman*, New Haven, CT: Yale University Press.

Simek, J. et al. (2012), "The Prehistoric Cave Art and Archaeology of Dunbar Cave, Montgomery County, Tennessee," *Journal of Cave and Karst Studies*, 74 (1): 19–32.

Simon, H. (1968), *Introduction to Printing: The Craft of Letterpress*, London and Boston, MA: Faber & Faber.

Situngkir, H. (2008), "The Computational Generative Patterns in Indonesian Batik." Available online: www.cogprints.org (accessed 12 June 2014).

Smith, L. (1994), *Modern Japanese Prints, 1912–89: Woodblocks and Stencils*, London: British Museum Press.

Somel, M., L. Xiling and P. Khaitovich (2013), "Human Brain Evolution: Transcripts, Metabolites and their Regulators," *Nature Reviews Neuroscience, 14*: 112–27.

Somervill, B. (2000), *Empire of the Aztecs: Great Empires of the Past*, New York: Chelsea House.

Sontag, S. (1995) (2008), "Against Interpretation," in *Art History and its Methods: A Critical Anthology*, 214–22, London and New York: Phaidon.

Spradley, J. (1979), *The Ethnographic Interview*, New York: Holt, Rinehart and Winston.

Stephen-Cran, J. (2009), *Awaken*, Glasgow: The Glasgow School of Art.

Strauss, A. and J. Corbin (1998), B*asics of Qualitative Research: Grounded Theory Procedures and Techniques*, 2nd edn, London: Sage.

Stringer, C. and C. Gamble (1993), *In Search of the Neanderthals: Solving the Puzzle of Human Origins*. London: Thames & Hudson.

Strohminger, N. and S. Nichols (2014), "The Essential Moral Self," *Cognition*, 131 (1): 159–71.

Sullivan, G. (2005), *Art Practice as Research: Inquiry in the Visual Arts*, Thousand Oaks, CA, and London: Sage.

Sullivan, G. (2008), "Methodological Dilemmas and the Possibility of Interpretation," *Working Papers in Art and Design*, 5. Available online: http://sitem.herts.ac.uk/artdes_research/papers/wpades/vol5/gdfull.html (accessed 29 May 2010).

Tarrant, N. (1987), "The Turkey Red Dyeing Industry in the Vale of Leven," in J. Butt and K. Ponting (eds), *Scottish Textile History*, 37–47, Aberdeen: Aberdeen University Press.

TechneGraphics (2007), "How Digital Printers Work." Available online: http://macgra.com/printips.html (accessed 4 December 2012).

Tedesco, L. (2000), "Lascaux (ca. 15,000BC)," in *Heilbrunn Timeline of Art History*, New York: The Metropolitan Museum of Art.

Tian, Y. (2007), "The Invention and Impact of Printing in Ancient China." Available online: http://educ.ubc.ca (accessed 21 February 2013).

Townsend, C. and P. Roberts (1999), *Into the Light: Photographic Printing Out of the Darkroom*, Bath: Royal Photographic Society.

Townsend, K. (2004), "Transferring Shape: A Simultaneous Approach to the Body, Cloth and Print for Textile and Garment Design (Synthesising CAD with Manual Methods)," PhD thesis, Nottingham Trent University.

Townsend, K., A. Briggs-Goode, and C. Northall (2010), "2D/3D/2D: A Diagnostic Approach to Textile and Fashion Research Practice," DUCK. Available online: http://www.lboro.ac.uk/departments/sota/research/Duck_NEW_2010/volume1.html (accessed 28 April 2011).

Treadaway, C. (2004), "Digital Reflection: The Integration of Digital Imaging Technology into the Creative Practice of Printed Surface Pattern and Textile Designers," *The Design Journal*, 7 (2): 3–17.

Treadaway, C. (2006), "Digital Imaging: Its Current and Future Influence upon the Creative Practice of Textile and Surface Pattern Designers," PhD Thesis, University of Wales Cardiff.

Treadaway, C. (2007), "Digital Crafting and Crafting the Digital," *The Design Journal*, 10 (2): 35–48.

Treadaway, C. (2009), "Materiality, Memory and Imagination: Using Empathy to Research Creativity," *Leonardo*, 42 (1): 231–7.

Treadaway, C. (2011), "Creative Momentum in the Information Age," essay in Momentum Exhibition Catalogue, Cardiff, Swansea Metropolitan University.

Treadaway, C. (2012), "Shorelines." Available online: http://www.cathytreadaway.com/research/shorelines (accessed 28 December 2012).

Turing, A. (1950), "Computing Machinery and Intelligence," *Mind*, 59: 433–60.

Tyler, D. (2005), "Textile Digital Printing Technologies," *Textile Progress*, 37 (4): 1–65.

Uglow, J. (2007), *Nature's Engraver: A Life of Thomas Bewick*, London: Faber & Faber.

Ujiie, H. (ed.) (2006), *Digital Printing of Textiles*, Cambridge: Woodhead Publishing.

Updike, D. (1962), *Printing Types: Their History, Forms, and Use: A Study in Survivals*, Cambridge, MA: The Belknap Press of Harvard University Press.

V&A (Victoria and Albert Museum) (2013), "Resist-Dyed Textiles." Available online: http://www.vam.ac.uk/content/articles/resist-dyed-textiles/ (accessed 12 March 2012).

Valentine, L. (2011), "Craft as a Form of Mindful Inquiry," *The Design Journal*, 14 (3): 283–306.

Veiteberg, J. (2005), *Craft in Transition*, trans. D. Ferguson, Bergen: Kunsthøgskolen i Bergen.

Virilio, P. (2007), *The Great Accelerator*, trans. J. Rose, Oxford: Berg.

Vogelsang-Eastwood, G. (2000), "Textiles," in P. Nicholson and I. Shaw (eds), *Ancient Egyptian Materials and Technology*. 275–9, Cambridge: Cambridge University Press.

Waite, M. (2009), "Sustainable Textiles: the Role of Bamboo and a Comparison of Bamboo Textile Properties," *Journal of Textile Apparel, Technology and Management*, 6 (2): 1–21.

Wakeling, C. (2013), *Letterpress: Gotthard de Beauclair, Monotype News Letter*, Oxford: The Fine Press Book Association.

Wallace, J. (2013), "Interaction Design, Heritage and the Self," *Interactions*, 20 (5): 16–20.

Wallis, S. (2010), "Toward a Science of Metatheory," *Integral Review*, 6 (3): 73–110.

Ward, C. et al. (2013), "Early Pleistocene Third Metacarpal from Kenya and the Evolution of Modern Human-like Hand Morphology," Proceedings of the National Academy of Sciences, 16 December. doi: 10.1073/pnas.1316014110.

Weber, W. (1966), *History of Lithography*, London: Thames & Hudson.

Weil, D. (1999), "A Course for Change," in J. Myerson (ed.), *Design Renaissance*, 119–24, Horsham: Open Eye Publishing.

Weiss, R. (2008), "Between the Material World and the Ghosts of Dreams: An Argument about Craft in Los Carpenteros," *The Journal of Modern Craft*, 1 (2): 255–70.

Whitehead, A. (1985), *Symbolism: Its Meaning and Effect*, New York: Fordham University Press.

Williams, J. (2012), *Type Matters!*, London, New York: Merrell.

Willig, C. (2008), *Introducing Qualitative Research in Psychology*, Maidenhead: Open University Press.

Winnicott, D. (1971), *Playing and Reality*, London: Tavistock.

Winnicott, D. (1973), *Playing and Reality*, Stuttgart: Klett.

Xin, J. (ed.) (2006), *Total Colour Management in Textiles*, Cambridge: Woodhead Publishing.

Yanagi, S. (1974), *The Unknown Craftsman*. Tokyo: Kodashana.

Ye, Y. (2000), "The Ozone Fading of Traditional Chinese Plant Dyes," *Journal of the American Institute for Conservation*, 39 (2): 250–7.

Yin, R. (1994), *Case Study Research: Design and Methods*, Thousand Oaks, CA, and London: SAGE.

Yoshida, H. (1939), *Japanese Woodblock Printing*, Tokyo and Osaka: The Sanseido Co.

Yunus, N. (2012), *Malaysian Batik: Reinventing a Tradition*, North Clarendon, VT, Tokyo and Singapore: Tuttle Publishing.

Yusoff, N. (2011), "A Study on Laser Cutting of Textiles. Laser Institute of America: Laser Applications and Safety." Available online: http://www.lia.org (accessed 8 September 2012).

Zachry, M. and C. Thralls, (2004), "An Interview with Edward R. Tufte," *Technical Communication Quarterly*, 13 (4): 447–62.

Zeisel, J. (1984), *Inquiry by Design: Tools for Environment-Behaviour Research*, Cambridge: Cambridge University Press.

INDEX

agave fibers 42
alginate 33, 87, 92
Althusser, L. 85–6
Aristotle 31, 39, 72
Arnheim, R. 40, 53
artists' instinctive way of working 96–7
authenticity 4, 78–80, 115
Awaken project 63

Bankes, H. 19
Banks, J. 76
Barber, E. 42
Barrett, E. 41
batik 26, 64–5, 70
Beckow, S. 41
Beddard, H. 52
Bell, Thomas 2, 18
Benjamin, W. 13–14, 52, 77–80
Biggs, M. 8
body art 12
Bolt, B. 86
Bowles, M. 59
Brandeis, S. 54, 59
Bremner, E. 63, 114
Briggs-Goode, A. 56
Britt, H. 63, 114
Bunce, G. 56
Bunnell, K. 63–4, 86

Calvino, I. 38
Campbell, J. 37, 59–60, 70, 92
Carden, Susan (author) 51–2, 64–6, 69–70, 74, 87, 108, 114

Carlisle, H. 3
case studies 4–5
cave painting 4, 8–12, 21–2, 28–33, 38, 42–7, 72–6, 80, 112
Centre for Advanced Textiles (CAT), Glasgow 51, 59, 61–3
Chalmers, Sylvia 61
Chapman, P. 84
childhood memories 68
China 14, 25
chintz production 15–16
Cie, C. 36
Clarkson, Thomas 18
closed-loop processes 56
cloth production 42
Collis, A. 37
color
    appreciation and application of 34
    composition of 42–3
colored images 25, 34
computers
    creative use of 32
    personal engagement with 52
continuous flows of ideas 38
cooperative endeavour 29
copies, making of 78
copperplate printing 10–11, 16–18, 41, 48, 112
copyright 60
cotton 25
    thick and *thin* 108
craft skills and craft techniques 86–92
creative knowledge possessed by designers 99
creative mindset 106

creative practice 84
creativity, characteristics of 76
crossing disciplines 109–10
Crump, Scott 20

dating of images 9
Day, Lucienne 61
Deleuze, G. 32–3
designer researchers 84
Dewey, J. 30–3, 42, 71, 113
digital images 40, 49, 96, 102
    characteristics of 109
"digital look" 113
digital and non-digital processes combined 82, 109
digital technology 2–4, 8, 60
digital textile printing 20–1, 59, 67, 73, 78
    changes brought about in designers' ways of
        working 96–106, 110, 113
    characteristics of 49, 54, 106–7
    constraints on the designer 47–9, 64, 96, 103,
        112–14
    criticisms of 99, 105
    cultural roots of 46
    definition and description of 38–9, 77
    differences from previous techniques 113
    different perspectives on 94
    essence of 80–1, 111
    generic qualities of 102–3
    impact of processes 107
    influences on 92
    knowledge and knowledge transfer in 72–7
    likely change in the role of 32
    loss of control in the design process 97, 100,
        102, 107
    neighboring disciplines 93–4
    origins of 82
    practice in action 107–9
    stages of 35–6
    transferrable components of 95–8
    used to re-interpret sketches 63
digitally printed images
    alternative methods for creation of 96
    characteristics of 80, 113–14
    subjective impressions given by 97, 105, 113
Disney Research 66
Dodds, D. 52
Dormer, P. 94
drawing by hand 106
Dreyfus, H. 74

Dutton, L. 42
dye sublimation printing 36
dyes 12–13, 16, 18, 26–7, 43

Earley, R. 56
Eckert, C. 64
Eco, U. 23, 30, 36, 81
ecological products 70
Egypt 13–14
Einstein, A. 7
emulation of analog techniques 101, 109
ethical considerations 70
experiential knowledge associated with digital
        printing 102–4
experimentation, scope for 109

fake artefacts 69, 99, 113
"fast fashion" 115
film archiving 104
Fiore, Q. 40
Flusser, V. 40, 79
Foucault, Michel 3, 40
founding process 1
"four causes" (Aristotle) 30–1, 72
Frayling, C. 32
Freedom of Creation (company) 20
Freeland, C. 44, 63
Friedman, K. 82, 85
fused deposition modeling (FDM) 66

Gale, C. 24
Gardner, H. 30, 36, 71, 76, 81
Glasgow School of Art 62–3
Goldsworthy, Kate 51, 56–60, 69–70, 114
Gotz, M. 68
Gray, C. 84, 94
Grossman, B. 20
grounded theory 65–6, 84
Guattari, F. 32–3
Gutenberg, Johannes 14

Hall, I. and D. 94
hand dyeing 88
handcrafting 54, 59, 67, 97, 102, 109–10, 112,
    115
handling of materials 52
hands, images of 111
Harrod, T. 63–4, 71, 85
Heidegger, M. 3, 23, 30–1, 36, 81, 85

Hohl, M. 85
Holler, A. 68
Hudson, S. 51–2, 66–8, 114
Hughes, B. 47, 113–14
Hyman, J. 41

IBM (company) 20
ikat effects 16, 65
Illustrator software 96
images
    perception of 40–1
    separation of 41
import restrictions 1
India 1, 13, 16, 25–6, 42, 60–1
Indonesia 26
inkjet printing 23–4, 27, 29, 36, 71–7, 87
internet sources for images 96

Jablonski, N. 12
Jacquard, J.M. 20
Japan 25, 63
Jaugeon, Jacques 17
Johnson, B. 27
Johnson, P. 8–9
Joseph, E. 36

Katrantzou, Mary 68–9
Kaur, J. 24
knowledge, categories of 3
Kyttanen, J. 40

laser cutting 19–20, 113
laser welding 56
Latour, B. 39
Leach, B. 44
letterpress 17
light used in digital printing 36–7
lithography 18–19

McCullough, M. 24, 52, 54
Mackinnon, Lana 61
Mackintosh, Charles Rennie 51–2, 62–3
McLuhan, M. 40, 44–5, 75
Malaysia 26
Malins, J. 84, 94
Margolin, V. 85
Marshall, J. 54–6
Marx, K. 31
mass production 37, 62, 78

materials
    designers' understanding of 104
    specific qualities of 101
Maxwell, Clerk 33–4, 41
memory *see* childhood memories
Merleau-Ponty, M. 79
metameric failure 27
Milliken (company) 20
mimicking processes 99–100, 103
Mirzoeff, N. 24
mordanting 13, 16–17
Morris, William 27
multiple materials 43
multiple processes 39

natural dyes 27
natural fibres 49
nettles 42
Nicol, K. 85
Niedderer, K. 84
Nimkulrat, N. 85–6

ontic entities 85
ontological positions 85
Oppenheim, B. 94
original designs or artefacts 77–8

paper manufacturing 14–15, 18
peer response to design practice 85
Peirce, C. 31
The People's Print project 59
perception 40–1
Perkin, William Henry 18, 26–7
Phillipson, M. 23
Philpott, R. 84
photography, uses of 56, 79
Photoshop software 96
pictograms 14
pigments 7–13, 21, 27, 29, 47, 111
Pike, A. 8–9
pixels 29, 52
Plato 31, 40
Polanyi, M. 30, 32
Pollock, G. 46
Polo, Marco 33
practical experimentation 87
practice-based and practice-led research 84–6
Prakesh, O. 16
preconditions 79

print heads 37
print technology 1–2, 14, 34; *see also* copperplate printing; digital printing; roller printing; screen-printing; three-dimensional printing; woodblock printing
procedural knowledge 84, 86
Pye, D. 23

Quinn, B. 64

Rancière, J. 78
rasterization 49–50
reflective practice and reflective practitioners 84
replication of analogue techniques 99–100
reproduction 77–8
    components and conditions of 79–80
resist dyeing 28
resist dyeing systems 26
resolution of a printed image 37
Rigley, Steve 51, 60–1, 69–70, 114
roller printing 18
Romaine de Roi type 21
rope making 24
Russell, Alex 51, 53–4, 114

Schön, D. 30, 46, 84
screen-printing 19, 27, 36, 41, 48, 62, 72, 96, 112
Scrivener, S. 30, 84
Senefelder, Alois 18–19
Sennett, R. 30, 35, 53
silk chiffon, *silk satin* and *silk dupion* 109
silkscreen printing 41
sketchbooks, designers' use of 106–7
skills of digital textile printing, mastery of 30, 35–6, 52–4, 71–2, 81–2, 86, 99, 102–5, 114–15
skills of practice-based research 86
skin decoration 11–13, 29
"slow design" 70
social engagements 112
Sontag, S. 52–3
spatial awareness 101–2
Spradley, J. 94
spray-painting 9–10, 21–2, 29, 112
stamping an image 1, 13, 112
stencils 9–12, 21–2, 25–8, 33, 65, 111–12

Stewart, Robert 51, 61–2
Stewart, Sheila 62
Stork (company) 21
studio practice and studio inquiry 4, 81–5, 96
Sullivan, G. 84
Sumatra 16
sustainability 56, 69
synthetic dyes 26–7, 43

tacit knowledge 30, 73–7
tactile visuality 100
tattooing 12, 112
technical knowledge, acquisition of 75–6
teddy bears 67–8
theory-building 82–3
three-dimensional (3D) printing 20, 51–2, 54
tie-dyeing 26
Timorous Beasties (company) 64
Townsend, K. 56
toxic waste 27
traditional skills 85–6
transferable knowledge linked to digital printing 102, 104–6
translation of printing techniques from one medium to another 100
Treadaway, Cathy 51, 53–6, 114
Turing, A. 24, 74
type, design of 17

Ujiie, H. 59
Updike, D. 17

Valentine, L. 86
vinegars 88
viscose 109

weaving 24–5
Weiss, B. 86
Wilson, J. 37
woodblock printing 13–18, 41, 112–13
wool yarn, use of 67–8
workmanship of certainty and *workmanship of risk* 98, 103–4

Yanagi, S. 23
Yoshida, H. 14